MW01492045

Monographs in Electrical and Electronic Engineering 18

Editors: E. H. Rhoderick, P. Hammond, and R. L. Grimsdale

Monographs in Electrical and Electronic Engineering

Permanent–Magnet and Brushless DC Motors

T. Kenjo

Professor in the Department of Electrical Engineering,
Institute of Vocational Training, Kanagawa, Japan

and

S. Nagamori

President and Chairman of the Board,
Nippon Densan Corporation, Kyoto, Japan

CLARENDON PRESS · OXFORD · 1985

Oxford University Press, Walton Street, Oxford OX2 6DP
Oxford New York Toronto
Delhi Bombay Calcutta Madras Karachi
Kuala Lumpur Singapore Hong Kong Tokyo
Nairobi Dar es Salaam Cape Town
Melbourne Auckland

and associated companies in
Beirut Berlin Ibadan Nicosia

Oxford is a trade mark of Oxford University Press

Published in the United States
by Oxford University Press, New York

© T. Kenjo and S. Nagamori, 1985

This English edition was revised and translated by T. Kenjo from
the Japanese book originally written with S. Nagamori and published by
Sogo Electronics Publishing Company, Tokyo

All rights reserved. No part of this publication may be reproduced,
stored in a retrieval system, or transmitted, in any form or by any means,
electronic, mechanical, photocopying, recording, or otherwise, without
the prior permission of Oxford University Press

British Library Cataloguing in Publication Data
Kenjo, Takashi
Permanent-magnet and brushless DC motors—
(Monographs in electrical and electronic engineering; 18)
1. Electric motors, Direct current
I. Title   II. Nagamori, S.   III. Series
621.46′2   TK2681
ISBN 0–19–856214–4
ISBN 0–19–856217–9 Pbk

Library of Congress Cataloging in Publication Data
Kenjo, Takashi.
Permanent magnet and brushless DC motors.
(Monographs in electrical and electronic engineering)
Translation of: Mekatoronikusu no tame no DC sābo mōta.
Includes index.
1. Electric motors, Direct Current.   Servomechanisms.
I. Nagamori, Shigenobu, 1944–   .   II. Title.
TK2681.K4613 1985   621.46′2   85–8862
ISBN 0–19–856214–4 (U.S.)
ISBN 0–19–856217–9 (U.S. : pbk.)

Filmset and printed in Northern Ireland at The Universities Press (Belfast) Ltd.

Preface

A great number of small electrical motors are manufactured for use in industrial robots, numerically controlled machines, computer peripherals such as floppy/hard disk drives and printers. The most significant feature of these machines is that their mechanical functions are controlled by several motors through electronic means. Since the mid-1970s in Japan, this technology has progressed rapidly and this engineering field is widely known as mechatronics.

Since 1977 I have written several books about various kinds of small electrical motors and their electronic controls. My fourth Japanese book, which deals with permanent-magnet and brushless DC motors, was entitled *DC servomotors for mechatronics* and published by Sogo Electronics Publishing Company. I wrote this in 1982 in collaboration with Mr S. Nagamori who had been one of my first students and later established his present company Nippon Densan Corporation in Kyoto. Since we had received an appreciative response from Japanese readers to this book, we decided to produce its English version to share our knowledge with world readers who are interested in this subject. It was, however, only at the beginning of 1984 that I started this work. After preparing a draft translation I revised it entirely, and added new parts to Chapter 8 to deal with PWM servo-amplifiers and an application of a microprocessor to a position control system. Furthermore I had many of the original illustrations improved.

I am pleased to acknowledge that this book was completed with the friendly support, help, and encouragement of several persons. Mr Nagamori financially supported me in producing the typescript on a wordprocessor, together with an assistant. Mr T. Kawawaki, manager of the Tokyo Branch of Oxford University Press, arranged for the publication of this book and always encouraged me. Mr T. Burger, a friend of mine, and Mr M. S. Lam, one of my present students, kindly helped me express my Japanese ideas in proper English. Mr T. Kato, who also used to be my student and now is with Nippon Densan Corporation, produced the illustrations. I wish to express my thanks to these persons. My thanks are also extended to the staff of the Clarendon Press who gave me helpful advice and refined my English.

Kanagawa, Japan   T. K.
September 1984

Contents

1. Principles of permanent-magnet DC motors

There are many different types of motors which operate on electro-magnetic principles. Of these, permanent–magnet DC motors have out-standing characteristics for use as the actuators in automated equipment. This chapter will focus on an explanation of the basic principles of permanent-magnet DC motors.

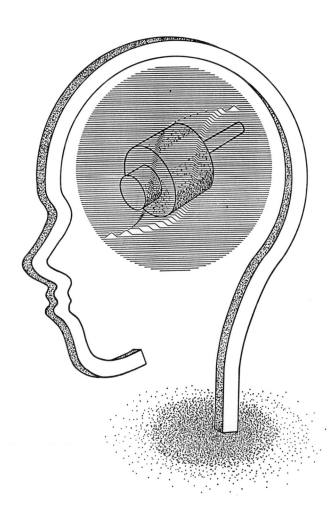

1.1   Explanation of basic terms

Although there are various types of direct current motors, the motor shown in Fig. 1.1 is suitable for understanding the basic principles.

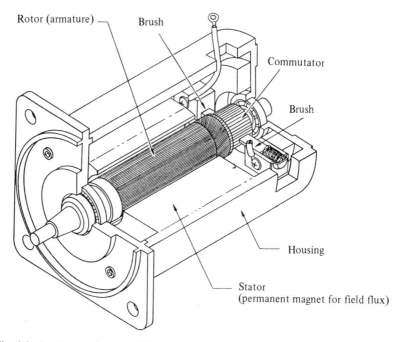

Fig. 1.1. A cutaway view of a DC motor.

However, it will be essential to recognize some basic terms before explaining these principles:

Rotor: The rotating part of the motor.

Stator: The stationary part of the motor.

Field system: The part of the motor which provides the magnetic flux needed for creating a torque. In Fig. 1.1 the field system consists of two permanent magnets and an iron housing, forming a part of the stator (see Fig. 1.2).

Armature: The part of the motor which carries the current that interacts with the field flux to create the torque. In the motor of Fig. 1.1 the rotor is referred to as the armature since it has coils wound around it. These coils serve to transfer the current from the brushes and commutator to the rotor.

Brushes: The part of the circuit through which the electrical current is supplied to the armature from a power supply. Brushes are made of graphite or precious metal. A DC motor has one or more pairs of

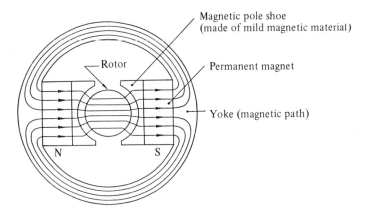

Fig. 1.2. Field system; movement of flux.

brushes. In Fig. 1.1 one brush is connected to the positive terminal of the power supply, and the other to the negative:

Commutator: The part which is in contact with the brushes. The current is properly distributed in the armature coils by means of the brushes and commutator.

1.2 Fleming's left-hand rule and generation of torque

In a DC motor the torque generation is based on Fleming's left-hand rule. In Fig. 1.3, a conductor is placed in a magnetic field and if current flows through the conductor, a force will act on it. The direction of the force is determined by the left hand rule as illustrated in the same figure, and the magnitude of the force is given by the following equation:

$$F = BIL \qquad\qquad (1.1)$$

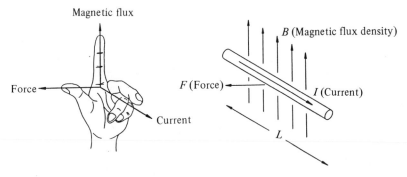

Fig. 1.3. Fleming's left-hand rule.

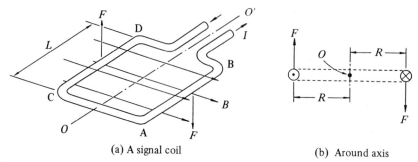

(a) A signal coil   (b) Around axis

Fig. 1.4. Coil in a magnetic field.

where B = magnetic flux density (unit: tesla T)

    I = electrical current (unit: ampere A)

    L = effective length of the conductor (unit: metre m)

    F = force (unit: newton N).

Figure 1.4 illustrates the torque obtained when a single-turn coil is placcd in a magnetic field. Here, there are two conductors present, $\overline{AB}$ and $\overline{CD}$. $\widehat{AC}$ and $\widehat{DB}$ are considered to be the connections between the two conductors and called coil-ends. The magnitude of the force acting on each conductor is determined by eqn (1.1). The directions of each force acting on coil pieces $\overline{AB}$ and $\overline{CD}$, respectively, are opposite to each other. In Fig. 1.4(b) the torque T around axis OO′ works clockwise and the magnitude is

$$T = 2RF = 2RBIL \qquad (1.2)$$

where T = torque (unit: newton metre, N m)

    R = distance from the centre to each conductor (unit: m)

The international unit (SI) for torque is N m. See Appendix for the unit conversions.

1.3 Torque constant

In the armature of the motor of Fig. 1.1, the current distribution is as illustrated in Fig. 1.5. If the current flowing in the conductors to the right of the symmetrical axis OO′ is in the direction of $\otimes$ (away from the reader), then current in the conductors to the left flows in the opposite direction of $\odot$ (towards the reader). The brushes and the commutators always distribute the direct current from the terminals into the rotor in the above manner.

In Fig. 1.5, the conductors in the right half are under the north pole and the conductors in the left half are under the south pole of the permanent magnet. (Actually, some of the conductors are not under a magnetic pole, but we will assume that they are under either north or south poles in order to simplify the discussion.) If we now assume that the

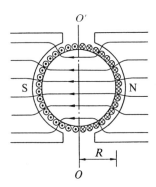

Fig. 1.5. Field flux and current distribution in the rotor.

magnetic flux density has an average value of B, then the torque $RBIL$ should work on every conductor and the whole torque T around the axis will be

$$T = ZRBLI$$
$$= ZRBLI_a/2 \qquad (1.3)$$

where Z is the total number of conductors and I_a is the current from the motor terminal which is equal to $2I$ (see Section 1.7).

In this model, the magnetic flux is equal to

$$\Phi = \pi RLB. \qquad (1.4)$$

Therefore, from eqn (1.3), we get

$$T = (Z/\pi)\Phi \cdot I_a/2. \qquad (1.5)$$

Now let us consider this equation. The number of conductors Z never changes in a finished motor. Because the magnetic flux Φ is determined by the motor dimensions and state of magnetization, $(Z/\pi)\Phi$ is a fixed value. Therefore we can conclude that the torque T is proportional to the armature current I_a. Now we define the torque constant K_T as:

$$K_T = (Z/2\pi)\Phi. \qquad (1.6)$$

Therefore, from eqn (1.5) we obtain

$$T = K_T \cdot I_a. \qquad (1.7)$$

Note that (N m A^{-1}) is used as the unit of torque constant.

1.4 Fleming's right-hand rule and back-e.m.f. constant

We have seen that a torque is generated when a current flows into the motor. In order to understand the relationship between the motor termi-

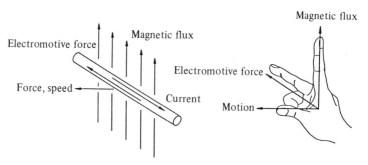

Fig. 1.6. Fleming's right-hand rule.

nal voltage and the current, and thus how the rotational speed is determined, we have to know how electrical power is generated inside the motor, as well as Fleming's right-hand rule, and the back-e.m.f. constant.

As illustrated in Fig. 1.6 a force works on the conductor and moves it at a speed V to the left. The conductor is moved by the action of the magnetic field and the current. Since the conductor now passes through the magnetic field, an electromotive force E is generated in the conductor. The magnitude of the force is:

$$E = vBL. \tag{1.8}$$

The direction is determined by Fleming's right-hand rule. That is, the direction in which the electrical power is generated is opposite to the direction of the current, so as to oppose its flow.

As each conductor passes through the north and the south magnetic poles, the electromotive force changes successively. But, as will be explained later, because of the brushes and commutator, the total electromotive force on each coil merges to the motor terminals. This voltage is the back electromotive force (back e.m.f.). The direction of this force is opposite to the terminal voltage applied. This value is directly proportional to the rotational speed Ω and is given by the following equation:

$$E = K_E \Omega. \tag{1.9}$$

The proportional constant K_E in this equation is the back-e.m.f. constant.

1.5   Relationships between torque and back-e.m.f. constant

We will now show how the back-e.m.f. constant K_E can be expressed in terms of other parameters. If the rotor is revolving at a speed of Ω radians per second, the speed of the conductor v is

$$v = \Omega R. \tag{1.10}$$

Therefore, the back electromotive force e generated in a conductor is

$$e = \Omega RBL. \tag{1.11}$$

If the total number of conductors is Z as explained in Section 1.7, then the number of conductors in a series connection is $Z/2$ and the total back-e.m.f. E at the motor terminals is

$$E = \Omega RBLZ/2. \tag{1.12}$$

By using eqn (1.4) we can express E in terms of the flux Φ as

$$E = (\Phi Z/2\pi)\Omega. \tag{1.13}$$

Therefore, by comparing with eqn (1.9), we obtain for the back-e.m.f. constant K_E:

$$K_E = (Z/2\pi)\Phi. \tag{1.14}$$

It is important to note that the torque constant K_T and the back-e.m.f. constant K_E are exactly the same as shown by eqns (1.6) and (1.14). It should be noted that K_T and K_E are only equal when a self-consistent unit system is used. The international system of units (SI) is one such system. For example, if K_T is equal to 0.05 N m A^{-1}, then K_E is equal to 0.05 V s rad^{-1}. The (SI) unit for the rotational speed is rad s^{-1}, but this unit is not always useful, because one revolution per second would be 6.28 rad s^{-1}. Conventionally, r.p.m. (revolutions per minute) has been used to measure the rotational speed of motors. The unit for the back-e.m.f. constant used in industry is V kr.p.m.$^{-1}$. When using the customary (but inconsistent) unit system, K_E can be determined by conversion if the torque constant K_T is known. In a similar manner, if K_T is measured, K_E can be found by conversion. Table 1.1 is the conversion table for these purposes. If a theory is developed under a consistent unit sytem, K_E and K_T could be the same:

$$K_T = K_E \equiv K \tag{1.15}$$

which can be called the motor constant.

Table 1.1. Unit conversions for torque constant and back-e.m.f. constant

Torque constant (K_T)		Back-e.m.f. constant (K_E)	
N m A^{-1}	oz in A^{-1}	V s rad^{-1}	V kr.p.m.$^{-1}$
1	141.6	1	104.7
0.098 07	13.89	0.098 07	10.27
0.007 061	1	0.007 061	0.7394
0.009 549	1.352	0.009 549	1

Note. For permanent-magnet and brushless DC motors, K_T and K_E have a specific relationship to each other so that the left side and the right side of the above table correspond to each other. Therefore, when K_T is 1 N m A^{-1}, K_E is automatically 1 V s rad^{-1}.

1.6 Stationary torque characteristics

We will now apply the knowledge gained in the previous sections to consider the relationship between the torque and the rotational speed of a DC motor.

A motor which uses permanent magnets to supply the field flux is represented by the simple equivalent circuit of Fig. 1.7. This is a series circuit of the armature resistance, R_a, and the back e.m.f., E.

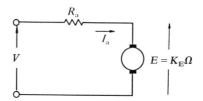

Fig. 1.7. Equivalent circuit of a DC motor.

If we ignore the voltage drop across the brushes, the equation for the voltage is

$$V = R_a I_a + K_E \Omega. \tag{1.16}$$

The armature current I_a is

$$I_a = (V - K_E \Omega)/R_a. \tag{1.17}$$

Therefore, from eqn (1.7), the torque T is

$$T = K_T I_a = \frac{K_T}{R_a}(V - K_E \Omega). \tag{1.18}$$

Table 1.2. The relationship between the supply and consumption of electric power in each action area of a DC motor

	Electric power supplied from power source	Joule heat / Current	Mechanical work per unit time
Motor action	VI_a Power source supplies electric power	$R_a I_a^2$ $I_a = (V - E)/R_a$	$EI_a = K_E \Omega I_a$ The motor works on the load
Generator action	$-V\|I_a\|$ Electric power is supplied to power source	$R_a I_a^2$ $I_a = (E - V)/R_a$	$-E\|I_a\| = -K_E \Omega \|I_a\|$ The load works on the motor
Brake action	VI_a Power source supplies electric power	$R_a I_a^2$ $I_a = (V + \|E\|)/R_a$	$-\|E\|I_a = -K_E \|\Omega\| I_a$ The load works on the motor

Note 1. Electric current I_c and rotational speed Ω are always regarded as positive here.
Note 2. Joule heat is always positive and generated in the armature winding in any case.

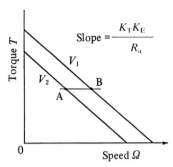

Fig. 1.8. Torque vs. speed characteristics of a DC motor.

Figure 1.8 shows the relation between T (torque) and Ω (rotational speed) at two different voltages. The torque decreases linearly as the speed increases. The slope of this function is a constant $K_T K_E / R_a$ and is independent of the terminal voltage and the speed. Such characteristics make the speed or position control of a DC motor easy. However, only DC and brushless DC motors have this feature—AC and stepping motors do not.

Explanation of terms

Starting torque: The torque when the motor is about to start, given by

$$T_S = K_T V / R_a. \tag{1.19}$$

No-load speed: The rotational speed when the motor has no load. If the bearings have no frictional load and no windage loss (electrical power consumed for driving the rotor against the air friction), then the relation will be given by the following equation:

$$\Omega_0 = V / K_E. \tag{1.20}$$

1.7   Functions of brush and commutator

As shown in Section 1.3, the brushes and commutator control the current distribution as illustrated in Fig. 1.5. At this point, we will only discuss the function and construction of the commutator. Figure 1.9 shows a typical commutator. The commutator segments are made of copper isolated with either mica or plastic. The risers are the parts to which the coil terminals are connected. The number of segments is the same as the number of coils, in general the higher this number, the more stable the commutator will be.

There are various methods of winding the coils and connecting the coil terminals to the risers. A lap-winding with nine coils is shown in Fig. 1.10. Part (a) is a schematic drawing of the winding and (b) is of the

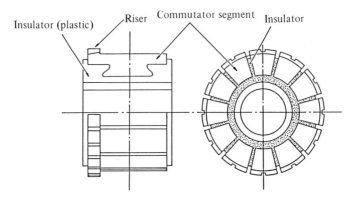

Fig. 1.9. Commutator structure.

connections. The important points in the winding and connections are as follows:

(1) The coil pieces of each coil (corresponding to the conductors described in Section 1.2) are assumed to be positioned at 180° to each other. In fact they are just slightly less than 180°, as illustrated in Fig. 1.11.

(2) As shown in Fig. 1.10(b), the coils are ring-connected to one another.

(3) The armature current I_a supplied from the positive brush terminal is divided into two circuits. Therefore, the relationship between the input current I_a and I flowing in each coil is

$$I = I_a/2. \tag{1.21}$$

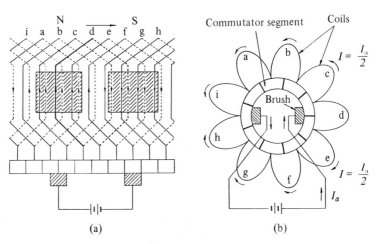

Fig. 1.10. Correlation of coil, brushes, commutator, and magnetic pole. (a) Schematic diagram of lap-winding, (b) coil connections.

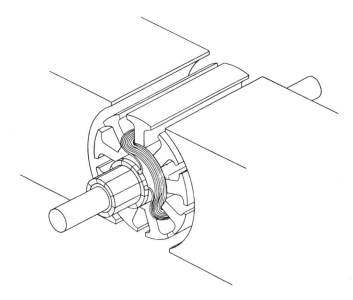

Fig. 1.11. Commutator segments and connection of coils. (This shows only one coil.)

(4) When there is a large number of coils and commutator segments, the number of series conductors is about half of the total number of conductors Z because the two circuits are in parallel.

(5) In Fig. 1.10(a) a current flows in the upward direction under the north pole, and a current flows in the downward direction in the coil pieces under the south pole. By reversing the polarity of the DC power source, the directions of the current and torque are reversed.

1.8 Pole zone and neutral zone

In Fig. 1.10(b), the coil d is now subject to commutation and must be placed in the neutral zone. If the rotor in Fig. 1.10(b) is rotating in a clockwise directiont, the current, having flowed up to this point in the counter-clockwise direction in coil d, is going to reverse. It should be noted that in a coil which is short-circuited by a brush the current is reversing its direction; this is referred to as 'the coil being subject to commutation' (see Fig. 1.12). A carbon brush is usually contacting two or more commutator segments, and one or more coils are short-circuited by the brush. When coil pieces pass through the magnetic field, the generated back e.m.f. causes a short-circuit current and produces the following adverse effects:

(1) Sparks will occur between the brush and the commutator.
(2) Heat and braking torque are produced.

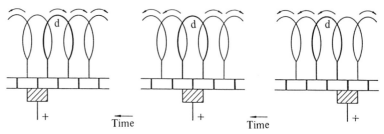

Fig. 1.12. Before and after coil d is subjected to commutation.

In order that a back e.m.f. is not generated in the coil under commutation, the conductors of this coil should not be in the magnetic field flux. The areas where no flux exists are called the neutral zones, and the areas where the field flux exists are called the pole zones (see Fig. 1.13).

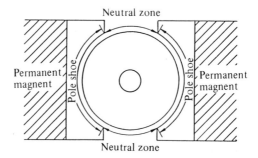

Fig. 1.13. Neutral zones and pole shoes.

1.9   Relationship between a motor and a generator

It is known that a DC motor can be used as a DC generator. Here we will study the relationship between a motor and a generator. It will also be explained that a generator called a tachogenerator or tachometer is used as a sensor of rotational speed and direction.

1.9.1   Simple generation

If a motor revolves due to an external force while leaving the motor terminal open, then by Fleming's right-hand rule, an electromotive force appears across the terminals. The magnitude is equal to the product of the rotational speed Ω and the back-e.m.f. constant K_E as stated in eqn (1.9). If an external resistance R_e is connected across the two terminals, a current caused by this electromotive force will flow. Because of the voltage drop caused by the armature resistance R_a, the terminal voltage V will be lower than the electromotive force. The value of V can be obtained from the equation:

$$V = \Omega K_E \frac{R_e}{R_e + R_a}.$$ (1.22)

This formula is valid in the case where a motor uses metallic brushes because of the almost negligible voltage drop across the brushes. In a motor which uses carbon brushes, V is lower than this value, due to the voltage drop across the brushes which is about 1 V.

A tachogenerator is a generator which detects rotational speed using the priciple of eqn (1.9) and has the following features:

(1) The output voltage is exactly proportional to the rotational speed, and even speeds less than one revolution per minute can be detected. However, it is designed to draw as little current as possible. Hence, the externally connected resistance must be as large as possible, i.e. above 1 kΩ.

(2) In order to decrease the voltage drop across the brushes, metallic carbon brushes which contain a large amount of silver dust are used.

(3) When the rotational direction reverses, the polarity of the output voltage reverses automatically. This makes the design of the control circuit simple. With an optical encoder, an electronic circuit is required to detect the rotational direction.

Figure 3.2 illustrates one example of a tachogenerator. A standard tachogenerator is designed so as to generate 7 V at a speed of 1000 r.p.m.

1.9.2 Regeneration

We will now consider the equivalent circuit of Fig. 1.7 to discuss the case in which the current returns to the power source from the motor as shown in Fig. 1.14. When the motor works with no-load at voltage V, the rotational speed is $\Omega_0 = V_0/K_E$. Next, when the motor is made to revolve by an external force at a speed greater than Ω_0, the back e.m.f. $E = K_E\Omega$ will be greater than the power source voltage V, and the current will return to the power source. Thus as the machine becomes a generator it will send the generated current back to the power source. This is called regeneration. At this point the current I_a is

$$I_a = \frac{E - V}{R_a}.$$ (1.23)

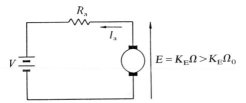

Fig. 1.14. Direction of current and magnitude of back e.m.f. in the regenerator action.

For generation action, mechanical work must be done by an external torque to rotate the machine faster than the normal speed. Work done in a unit of time is converted to the power P_0:

$$P_0 = EI_a = K_E \Omega I_a \tag{1.24}$$

and a part of this is consumed as Joule heat P_L:

$$P_L = R_a I_a^2, \tag{1.25}$$

and the remainder P:

$$P = P_0 - P_L = K_E \Omega I_a - R_a I_a^2 \tag{1.26}$$

is the regenerated power. From eqn (1.23) we get

$$P = VI_a. \tag{1.27}$$

Regeneration will occur in the following two cases:

(1) As mentioned above, when the machine works at a constant voltage, regeneration will occur if the machine is rotated faster than the no-load speed at that voltage.

(2) When the motor rotates at voltage V_1 and this voltage is lowered to voltage V_2, the machine will be brought into the regeneration action until the speed decreases to the no-load speed V_2/K_E.

1.9.3 *Machine action as a brake*

When a large external counter-clockwise torque is applied to the shaft of the motor which is now trying to revolve in a clockwise direction, the motor will act as a brake. Since in this case the direction of rotation is opposite to the natural direction, the polarity of the back-e.m.f. is reversed and the same as that of the terminal voltage (see Fig. 1.15). Such a situation causes a larger current to flow. The magnitude of the current is given by

$$I_a = \frac{V + |E|}{R_a}. \tag{1.28}$$

When a terminal voltage at which the machine has been operating as a motor is reversed, the machine will operate as a brake until the rotational

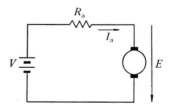

Fig. 1.15. Equivalent circuit when the motor is used as a brake.

direction reverses. This is often used to decelerate a motor effectively, but we must be careful with demagnetization due to a large current (see Section 2.4).

1.9.4 *Relationship between motor, generator, and brake*

The natural operational state of the working motor is called the electrical motor action. Other states include the generation action and the brake action. The relationships between these three are illustrated in Fig. 1.16.

These relationships can also be stated as follows:

The motor action: In the speed range between zero and the no-load speed V/K_E, the machine operates as a motor at a speed which depends on the load.

The generator action: The torque is negative in the speed range greater than the no-load speed V/K_E. Since the rotational speed is opposite to the direction of the torque, the machine is acting as a kind of brake. This state is also called a regenerative brake because the machine generates electrical power and returns this power to the source. The significance of the regenerator action can be well explained by using the example of an electric vehicle. When an electric car, with an engine that is either a permanent magnet DC motor or a similar DC motor, travels down a slope, we can apply to the engine an appropriate voltage at which the no-load speed is a little lower than the car's speed. In this instance the engine becomes a regenerative brake. Since the torque of the engine is negative, the engine speed does not increase much. The gravitational

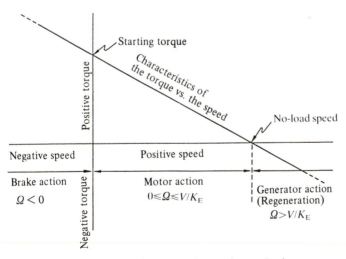

Fig. 1.16. The relations between speed ranges and operation modes (motor, generator, and brake) when a DC machine is operated at a constant voltage V; the torque vs. speed characteristics are given by a straight line through the entire range.

potential energy is converted to electrical power through regeneration, and turned into chemical energy by charging the battery. This is the best braking method for a DC motor, but it is not easy to make a circuit to efficiently utilize this method.

The braking action: The braking action is produced when the motor is made to revolve against its natural direction. This is the easiest way to brake, but a serious problem arises in the large amount of heat generated in the motor.

1.10   Conversion of energy through a DC machine

Let us consider a permanent-magnet DC machine from the viewpoint of energy conversion. Conversion of energy undertaken through a permanent-magnet DC machine is illustrated in Fig. 1.17. All three cases are expressed by the following equation:

$$VI_a = (E + R_a I_a)I_a = EI_a + R_a I_a^2 \qquad (1.29)$$

where,

VI_a is the electrical power supplied from the power source,

EI_a is the work done per unit time to supply current I_a against the back e.m.f., or the mechanical output power, and

$R_a I_a^2$ is the Joule heat.

Note that here the positive direction of the terminal voltage, the back e.m.f., and the current are taken as shown in Fig. 1.7. The current I_a is again:

$$I_a = (V - E)/R_a. \qquad (1.30)$$

However, the absolute value of I_a is used in Table 1.1.

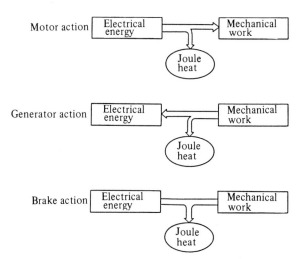

Fig. 1.17. The conversion and flow of energy in a DC motor.

The details of the meaning of EI_a are as follows: I_a flows against the back e.m.f. E which opposes it. Therefore, EI_a is the work done per unit time on the armature. A part of this work may be dissipated into heat but is ignored here. Hence the whole work done electrically is considered to be converted to mechanical work which is done on the load by the rotor. (In Chapter 6 the heat loss will be discussed.)

Because in the generator action E is higher than V and the current direction is reversed according to eqn (1.29), VI_a and EI_a become negative. The physical explanations for these terms are as follows:

Negative VI_a: Electrical power is supplied to power source.

Negative EI_a: Work is done on the rotor (armature) by the load. An external torque works on the rotor making it revolve.

For the brake action, E becomes negative because the direction of rotation is negative. In this case the above explanation of negative EI_a is also applied. In the brake action, the power source supplies energy and the work is done on the rotor by an external force. As both kinds of energy are dissipated into the Joule heat in the armature winding, the increase in temperature is high.

2. Structure of permanent-magnet DC motors

There are many different types of permanent-magnet DC motors. They can be classified according to their field systems, armature structures, and arrangement of brushes and commutator. An explanation of DC motors from this perspective will be given in this chapter. The armature structures of moving-coil motors will be discussed in detail in the next chapter.

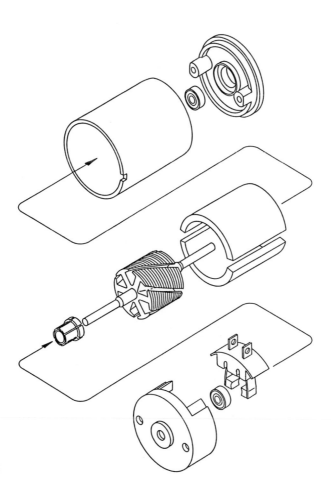

2.1 Fundamentals of permanent magnets

The structure of the field system is closely related to the nature of the permanent magnets used. Figure 2.1(a) shows a hysteresis loop. The reader is assumed to have some knowledge of magnetism and the hysteresis loop. In many cases, a permanent magnet is used in the second or fourth quadrant of the hysteresis loop. In these quadrants the directions of the magnetic field intensity H and the magnetic induction B oppose each other.

The second quadrant of the B–H curve is called the demagnetization curve. Magnetic characteristics in this region are called the demagnetization characteristics. The main points concerning the demagnetization characteristics are as follows:

(1) *Magnetic remanence and coercive force.* When a permanent magnet has been magnetized once, it will remain magnetized even if its magnetic field intensity is decreased to zero. The magnetic flux density at this point is called the magnetic remanence and designated by B_r. Furthermore, when the magnetic field intensity is increased in the opposite polarity along the demagnetization curve, the flux density will eventually become zero. Here, this field intensity is called the magnetic coercive force or coercivity and designated by H_c.

(2) *Energy product and maximum energy product.* The absolute value of the product of the flux density B and the field intensity H at each point

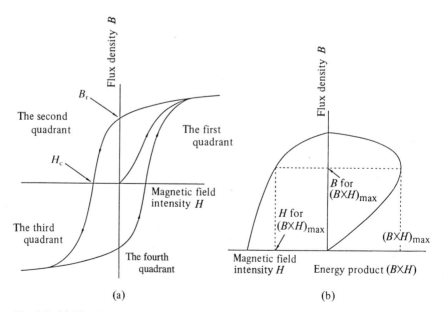

Fig. 2.1. (a) The hysteresis loop, and (b) demagnetization curve and energy product.

along the demagnetization curve is called the energy product. Figure 2.1(b) shows the energy product as a function of B in the second quadrant. The maximum value of the energy product is called the maximum energy product, and this quantity is one of the indexes of the strength of the permanent magnet.

Units

The SI unit for magnetic field intensity and magnetic coercive force is $A\,m^{-1}$, but physicists use Oersted (Oe) as their unit. The conversion between these units are given by

$$1\,A\,m^{-1} = 4 \times 10^{-3} = 1.256 \times 10^{-2}\,Oe \qquad (2.1)$$

$$1\,Oe = 1/4 \times 10^{3} = 79.577\,A\,m^{-1}. \qquad (2.2)$$

The SI unit for magnetic flux density and magnetic remanence is Tesla (T) and $1\,T$ equals $1\,Wb\,m^{-1}$ of the m.k.s. unit. Physicists use Gauss as their unit:

$$1\,T = 10\,000\,\text{Gauss},$$

$$1\,\text{Gauss} = 0.0001\,T.$$

The SI unit for the energy product is $J\,m^{-3}$, but $kJ\,m^{-3}$ is usually used.

(3) *Recoil lines.* A minor loop in the second quadrant of the hysteresis loop may be approximated by a straight line and called a recoil line. The line AC in Fig. 2.2 is a recoil line.

(4) *Stabilization.* The torque constant and back-e.m.f. constant may decrease if the field flux level becomes low while the motor is working. This is called demagnetization. To avoid demagnetization during operation it is necessary to stabilize the field magnets before using them by locating the operating point, which is the coordinate point specified by the present flux density B and the field intensity H, on an appropriate recoil line. This process is called stabilization and is usually done by motor manufacturers.

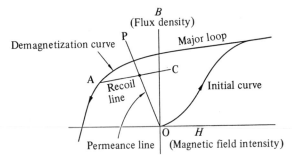

Fig. 2.2. Recoil line.

(5) *Operating point.* In a DC motor which is not supplied with a current, the *B–H* state of the permanent magnet is located at the intersection of the recoil line and the permeance line OP. The permeance line is determined from the characteristic quantities of the machine structure: i.e. the gap length, magnetic path length, and number of turns of the coils. During operation, however, the operating point is biased by the armature current. At one edge of a magnet the operating point is transferred to one side of the recoil line, and at the other edge of the magnet it is transferred to the other side of the recoil line. Hence as a whole, the magnetic flux supplied by the magnet remains almost constant unless the operating point deviates to the major demagnetization curve.

2.2   Kinds of permanent magnet

There are basically three different types of permanent magnet which are used in small DC motors:

    (i) Alnico magnet,
    (ii) Ferrite or ceramic magnet, and
    (iii) Rare-earth magnet (samarium–cobalt magnet).

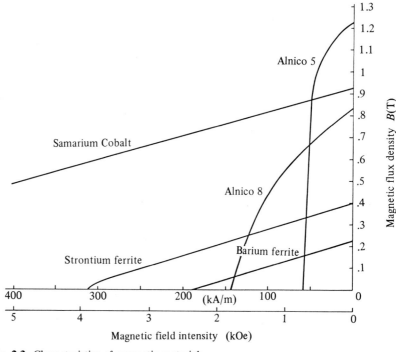

Fig. 2.3. Characteristics of magnetic materials.

The *B–H* characteristics of these three types vary greatly as shown in Fig. 2.3. Their features are as follows:

(1) *Alnico magnet.* This magnet provides a very high flux density but a low coercive force. When the coercivity is low and two opposing magnetic poles locate at a close distance, the poles can weaken each other. Therefore an Alnico magnet is used after being magnetized lengthwise.

(2) *Ferrite magnet (ceramic magnet).* Unlike an Alnico magnet, this magnet has a low flux density but a high coercive force. It is therefore possible to magnetize it across its width because of its high coercive force. Ferrite magnets are widely used because the material and production costs are both low.

(3) *Samarium–cobalt magnet.* This is a rare-earth magnet which has both high magnetic remanence and high coercive force. Since the initial cost is high, this magnet was first used in servomotors for aircrafts and military equipment as well as in motors used in computer peripherals. Since then, it has spread gradually into wider areas of use.

2.3 Structure of field system

Since the structure of the field system is very important in the use of the magnet, case studies are dealt with here for each type of magnet.

(1) *Alnico magnet.* Since Alnico magnets have a B_r, they are used in high-grade DC servomotors. To achieve a high magnetic flux they must be magnetized lengthwise for the reason stated above. See Fig. 2.4 which illustrates some examples of the field system. Figures 2.4(a) and (c) are of two-pole structures; (a) must have a non-magnetic material as its housing and therefore aluminium or its alloys are used, and (c) uses a mild steel housing to serve as a path for magnetic flux. The constructions in Figs. 2.4(b) and (d) are of four-pole motors; (d) uses a cylindrical housing of mild steel for the yoke or flux path, while (b) uses four pole shoes, and needs a non-magnetic housing or another means to fasten the magnets and pole shoes. Figure 2.5 shows a two-pole micromotor using an anisotropic Alnico magnet.

It is obvious from these figures that the number of magnetic poles is always even. In permanent-magnet DC motors it is usually two or four, but some motors have six or more magnetic poles.

(2) *Ferrite magnets.* Because ferrite magnets have a high coercive force, they do not demagnetize themselves even if the magnetization is widthwise. Figure 2.6 shows cross-sectional schemes of the field systems which use one or two ferrite magnets. Since ferrite magnets have a low flux density, the arrangement in Fig. 2.7 is often employed; the magnet is longer than the rotor to produce a high flux density in the armature. Since in this arrangement the area of mild steel housing is large, it is possible to make it thin so as to reduce the motor weight.

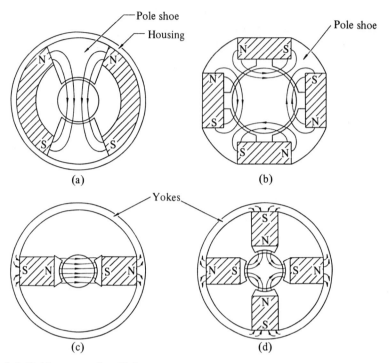

Fig. 2.4. Field system using Alnico magnets.

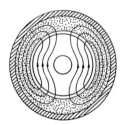

Fig. 2.5. Field construction in a micromotor using an Alnico magnet.

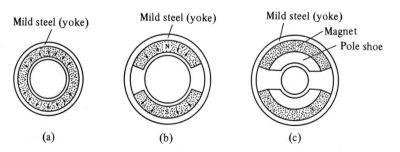

Fig. 2.6. Field construction using ferrite magnets. (a) Ringed anisotropic ferrite magnet, (b) anisotropic ferrite magnet in a dovetail shape, and (c) with dovetail pole shoes.

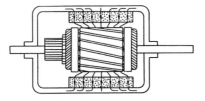

Fig. 2.7. Longitudinal cutaway view of ferrite magnets in a DC motor.

(3) *Rare-earth magnets.* Since this type of magnet has a high coercive force, it can be magnetized widthwise like a ferrite magnet. Because of high material cost, rare-earth magnets are made as thin as possible. The field system arrangements using rare-earth magnets are similar to those employed for ferrite magnets as shown in Fig. 2.8. However, since samarium–cobalt magnets are of high remanence, the rotor designs are different from those of motors with ferrite magnets. With an equal area and equal length magnet, the flux density is twice that of the ferrite magnet.

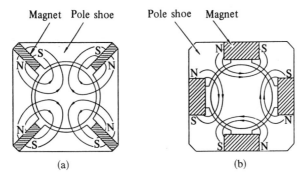

Fig. 2.8. Comparison of (a) a field system using samarium–cobalt magnets and (b) the one using Alnico magnets.

Hence the motor torque would show a similar increase, but this is not always the case, because a ferrite motor can increase its electrical loading by the adjustment of the slot/tooth widths as will be discussed in Section 2.5.2. An approximate comparison of a samarium–cobalt motor to an equal-size ferrite motor is as follows:[1]

Torque	approx. 1.5 times,
Mechanical time constant	approx. 0.5 times,
Peak output power	approx. 2.0 times,
Electrical time constant	approx. 0.7 times.

2.4 Demagnetization and counter-measures

In permanent-magnet DC motors, the flux produced by the field magnet is sometimes decreased due to the magnetic field created by the armature current. This adverse effect is known as demagnetization. The mechanism of demagnetization and its counter-measures will be discussed in the following section.

2.4.1   *Process of demagnetization*

Figures 2.9(a) and (b) show the flux distribution due to the field magnet and that due to the armature current in a two-pole motor, while Fig. 2.9(c) shows the combined flux distribution. In the armature core the field flux and the armature flux intersect at right angles. In the airgap, the flux density increases at one end of the pole and decreases at the other end. Demagnetization occurs when the flux density falls below the lower limit. This is shown in Fig. 2.10.

As explained earlier, permanent-magnet DC motors should be stabilized before they are shipped from the manufacturer. Stabilization means treatment of the magnet so that the magnetized state never changes during operation. This is done by setting the $B-H$ relation on a recoil line. As previously stated, when no armature current is supplied, the operating point stays at the intersection Q of the recoil line AC and the permeance line OP. The permeance line is a straight line with a

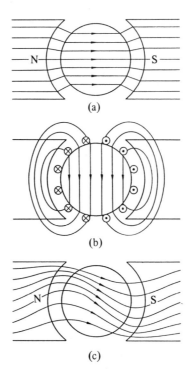

(a)

(b)

(c)

Fig. 2.9. Flux distribution inside a DC motor. (a) Flux produced by a field magnet, (b) flux produced by armature current, and (c) combined flux.

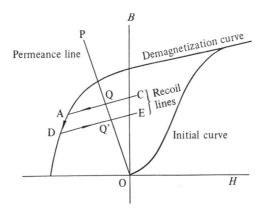

Fig. 2.10. Demagnetization curve, recoil lines, and permeance line.

negative slope and passes through the origin. The slope is determined by the construction parameters of the motor. At the intersection Q, the requirements of the $B-H$ relation both by the recoil line AC and by the permeance are satisfied. As shown in Fig. 2.9, when a large current flows in the armature, the magnetic field is strengthened at one end of the pole and weakened at the other. The problem occurs at the weakened side. When the field due to the armature is too strong, the operating point goes over the end of the recoil line, moves into the major hysteresis loop, and finally reaches point D. When the armature current decreases, the operating point will move along the new recoil line DE and end up at point Q'. Consequently, the flux density decreases. This process is known as demagnetization.

As stated in Section 1.9, the armature current is increased the most when the motor is decelerated by inverting the terminal voltage. At the moment when the terminal voltage is inverted, the current is roughly twice the starting current or the current the motor consumes at starting. Therefore, there may be a possibility of demagnetization at one side of the pole due to this high current. When the terminal voltage is inverted again, the other ends of the poles are also demagnetized; as a result the torque constant and back-e.m.f. constant are both reduced, and the motor performance decreases. By measuring the no-load speed, we can detect demagnetization. Since the no-load speed equals the terminal voltage divided by the back-e.m.f. constant as seen in Fig. 1.8, the no-load speed increases when demagnetization occurs.

2.4.2 Limiting demagnetization

There are several ways to limit demagnetization. One way is to keep the current below the maximum. Another way is by the use of pole shoes as shown in Fig. 2.11. The pole shoe is the mild-steel part which is attached

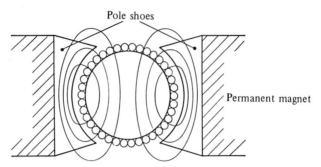

Fig. 2.11. Armature flux passing through pole shoes of soft steel.

to a permanent magnet to collect the flux and then transfer it to the air gap. As shown in Fig. 2.11, the flux produced by the armature current passes through the pole shoe. This flux does not affect the permanent magnet. Pole shoes are useful also for collecting flux to achieve a higher flux level in the air gap.

For a moving-coil or slotless armature, the pole shoes may be of solid mild steel. However, for a slotted armature, laminated steel sheets are used because strong eddy currents can be produced in solid steel due to the slot pulsation of the flux, and heat is produced. When the motor speed is high, the eddy-current loss in the laminated pole shoes is high, which is therefore the main disadvantage.

The samarium–cobalt magnet has very high coercive force, and the recoil line almost coincides with the demagnetization curve, which means that demagnetization does not occur.

Figure 2.12 shows an example of a field structure having no pole shoes. In this motor, the air gaps at the end of the permanent magnets are made slightly wider in order to minimize the flux produced by the armature current.

Fig. 2.12. Demagnetization is minimized by slightly widening the air gaps at the pole edges.

2.5 Armature structure

There are three fundamental armature structures shown in cross-section in Figs. 2.13, 2.14, and 2.15, respectively for slotted rotor, slotless rotor, and moving-coil rotor.

2.5.1 *Slotted rotor*

The core of a slotted rotor is a lamination of silicon steel sheet or carbon steel sheet, and windings are built up in the slots. In the slotted rotor, the torque acts directly on the durable iron core, not on the fragile coils. Thus a slotted rotor is more durable, and therefore has been in use for a long time. A core having many slots is usually desirable, because the greater the number of slots, the less the torque ripple, called cogging, and electrical noise. Cores having even numbers of slots are often used for the motors manufactured by an automated mass-production process because of the ease of production. But cores with odd-numbered slots are preferred because of low cogging torque.

As seen in Fig. 2.13, twisting the rotor reduces the cogging torque that is produced by the interaction between the rotor teeth and the pole shoes. This is called skewing.

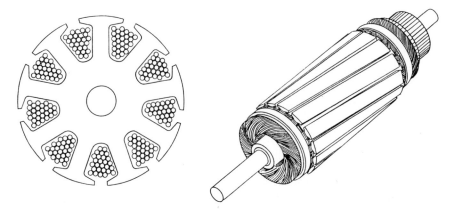

Fig. 2.13. Slotted rotor.

2.5.2 *Iron machine and copper machine*

There are two general ideas concerning slotted motor design, namely the iron machine and the copper machine. These concepts are briefly explained here.

The torque given by eqn (1.5) is rewritten as:

$$T = (ZI_a) \cdot \Phi/2\pi. \qquad (2.3)$$

Here, ZI_a, the product of the total number of conductors and the current, is related to the copper wire in the slots and is called 'electrical loading'. On the other hand, Φ is the field flux which passes through the core. Hence this quantity is related to the thickness of the core teeth, and called 'magnetic loading' (see Fig. 2.16).

When the rotor size is fixed, wide teeth must be used in order to get as much flux as possible. Then because the room for ZI decreases, the

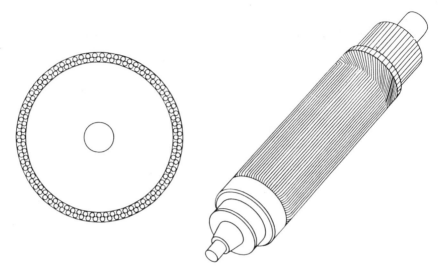

Fig. 2.14. Slotless rotor.

number of turns must be reduced or thin coils must be used, which will inevitably increase the Joule heat. On the other hand, ZI can be increased by making the teeth thin, but in this case it is difficult to get a high flux level.

In the case of a ferrite-magnet motor, a core with thin teeth and larger slots is used because the flux density provided by the ferrite magnet is low. Since the slots are large here, more copper wire can be installed in them and such a motor is often referred to as a 'copper machine'. On the other hand, a motor using Alnico or rare-earth magnets employs an iron

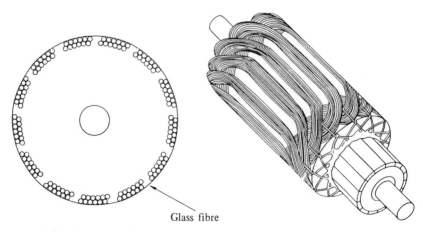

Glass fibre

Fig. 2.15. Moving-coil rotor.

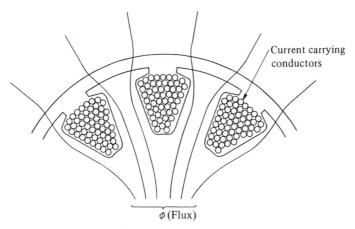

Fig. 2.16. Flux and current distributions.

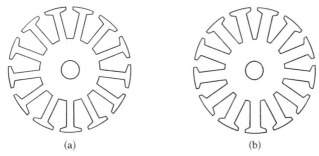

(a)   (b)

Fig. 2.17. Iron cores for (a) copper machine and (b) iron machine.

core with wider teeth and narrower slots as shown in Fig. 2.17(b) and it is referred to as an 'iron machine'.

When ferrite magnets are used, there seems to be a lot of wire packed in each of the larger slots, but this is not always the case. Figure 2.18 shows a typical way in which the wires are put in the bottom of the slots.

Fig. 2.18. Winding which reduces the rotor's moment of inertia.

This makes the rotor inertia small, which will then result in a small mechanical time constant.

2.5.3 Slotless rotor

Extremely low cogging torque can be produced by fixing the winding on a cylindrical iron core without using slots. In this case the torque is exerted directly on the wires (coil pieces) due to Fleming's left-hand rule. However, it is likely that the flux will decrease since the gap between the rotor core and the pole shoes is large. Therefore, rare-earth magnets or large, long Alnico magnets must be used to get sufficient flux.

2.5.4 Moving-coil rotors

In order for the rotor to have the smallest moment of inertia, a heavy iron core should not be used in it. In recent years, this type of rotor has been used in various applications. Since in a moving-coil motor all iron cores are stationary, i.e. they do not move in the field flux, no eddy-current or hysteresis losses are produced in them. For this reason the efficiency of a moving-coil motor is high in high speed ranges. This will be explained further in the next chapter.

2.6 Brushes and commutator

Brushes are important components of a DC motor, and they are made of graphite or precious metal.

2.6.1 Graphite brushes and commutator surface

A standard commutator segment is made of copper. The brush is made of graphite by the process of compression moulding and high-temperature calcining. The commutator segments are insulated with mica or plastic (see Fig. 2.19). The commutator surface can be rubbed off through sliding

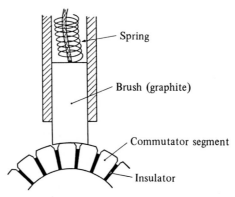

Fig. 2.19. Brush and commutator.

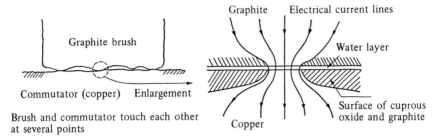

Fig. 2.20. Contacting area of a brush and a commutator segment; they are in contact at several points.

contact with the brushes. Since copper is rubbed off faster than mica, mica is usually embedded in the commutator surface.

When the commutator surface is covered with cuprous oxide (Cu_2O) several microns thick, mechanical wear is decreased and the life span is increased. This type of surface gives the commutator a lustrous dark brown colour. The moisture in the air is decomposed by electrolysis at a current density of 5–8 A cm^{-1} during operation. The atomic oxygen thus produced and the copper of the commutator surface chemically combine to form a new cuprous oxide film. Thus, mechanical wear and reproduction of the cuprous oxide is equilibrated. The cuprous oxide film acts as a kind of insulator, and it is believed that the current flows by breaking through the insulator film at several points which are in contact with the brushes (see Fig. 2.20). There is a voltage drop of 1–1.5 V between the brush and the commutator (see Fig. 2.21). This potential drop, which is

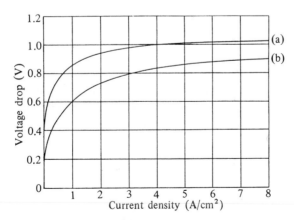

Fig. 2.21. Voltage drop across a brush and a commutator segment as functions of current. (a) Current flowing from commutator to brush, and (b) current flowing from brush to commutator.

known as a contact potential difference, is desirable for commutation because the electromotive force generated in the coil, subject to commutation, is absorbed by the contact potential difference, which results in the suppression of sparks. However, for low voltage motors, such a potential drop lowers their efficiency. Metallic graphite brushes mixed with copper or silver dust yield a low potential drop, and therefore they are used for motors which are operated on a low voltage with a high current. However, the quality of commutation decreases with these types of brushes; more sparks are produced and mechanical wear of the brushes is greater. For tachogenerators which are DC generators used as a speed sensor, brushes containing a high percentage of silver are used.

Figure 2.22 illustrates graphite (carbon) brushes; (a) shows the brushes for the printed motor, and (b) a micromotor. These brushes use a metallic as well as a rubber spring thus adding pressure to the brush and absorbing oscillations.

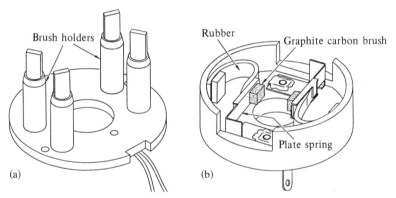

Fig. 2.22. Holders of carbon (graphite) brushes (a) for a printed motor, and (b) for a micromotor.

2.6.2 *Metallic brushes*

The size of the graphite brushes is a problem for micromotors of less than 10 W. The voltage drop across the brushes decreases the energy conversion efficiency of the motor as stated before. Therefore, metallic brushes are used in high-grade micromotors (see Fig. 2.23). Since metal contact is undesirable for stable commutation, selection of the metals was once a problem, but, now precious metals (silver, gold, platinum, and palladium) are used.

The metallic brushes oscillate at their resonant frequencies during revolution, and repeatedly come in and out of contact with the commutator. To solve this problem, the brush is divided into three sections

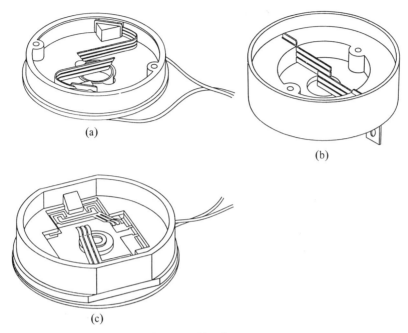

Fig. 2.23. Arrangements of precious metal brushes.

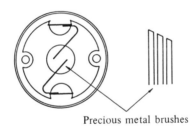

Precious metal brushes

Fig. 2.24. Brush is split into three pieces.

with different resonant frequencies as shown in Fig. 2.24. When such brushes are used, at least one of these three sections is in contact with the commutator at all times. Figure 2.23(a) shows the diagonal direction of the brushes in Fig. 2.24.

2.7 Examples of the parameters

Tables 2.1 and 2.2 show the parameters for slotted and slotless motors in practical use. Chapters 6 and 7 explain the meanings of several of these parameters. These tables are made by re-editing manufacturers' data. The

Table 2.1. Catalogue data for slotted motors

Manufacturers		Yasukawa Electric Mfg. Co., Ltd.						Tamagawa Seiki Co., Ltd.					
Item		UGJMED -10M	UGJMED -40M	UGJMED -40L	UGTMEM -01SB4	UGTMEM -03MB2	UGTMEM -06SB2	TS908N7 -E4	TS908N8 -E3	TS688N6 -E3	TS902N2 -E6	TS668N4 -E6	TS906N2 -E13
Inertia J	$10^{-6}\,\mathrm{kg\,m^2}$	600	1600	2000	1.57	23.5	95	0.918	1.57	2.50	28.4	39.2	234
Electrical time constant τ_E	ms	6	12.2	10.1	0.3	0.8	1.9	0.45	0.3	0.8	1.6	1.3	1.4
Mechanical time constant τ_M	ms	13.7	24.8	28.5	4.1	6.5	11.7	9	7	8	12	12	18
Torque constant K_T	$10^{-2}\,\mathrm{N\,m\,A^{-1}}$	47	37	50.6	3.4	7.5	9.3	3.82	3.92	3.72	6.43	6.47	12.5
Back-e.m.f. constant K_E	$10^{-2}\,\mathrm{V\,s\,rad^{-1}}$	47	37	50.6	3.4	7.5	9.3	3.82	3.92	3.72	6.43	6.47	12.5
Armature resistance R_a	Ω	5.0	1.05	1.3	3.2	1.59	1.02	14.3	6.9	4	1.7	1.3	1.05
Power rate	$\mathrm{kWs^{-1}}$	1.5	1.5	2.9	1.5	2.4	1.6	0.20	0.71	0.35	0.87	1.92	1.73
Rated continuous torque T	$10^{-1}\,\mathrm{N\,m^{-1}}$	0.95	1.53	2.40	0.05	0.24	0.39	0.137	0.333	0.294	1.57	1.96	6.38
Rated rotational speed Ω	r.p.m.	1000	1000	1000	3000	2000	1300	3750	3000	3300	4000	4000	3000
Rated output P_0	W	100	160	250	15	50	53	5	10	10	60	80	200
Rated voltage	V	64	44	60	20.3	24.4	19.8	21	21	18.3	30.8	31.3	43
Weight	kg	6	10.5	12	0.22	1.1	1.6	0.09	0.15	0.4	1.3	1.5	3.0

This table is re-edited from manufacturers' catalogues

Table 2.2. Catalogue data for slotless motors

Item		Yasukawa Electric Mfg. Co., Ltd.			Olympus Opto Electronics Co., Ltd.			
Manufacturers		UGMMEM-06AA1	UGMMEM-13AA-	UGMMEM-25AA1	OMS-312	OMS-512	OMS-1024	OMS-2024
Inertia J	10^{-6} kg m^2	56.7	141	283	0.13	0.22	0.58	1.73
Electrical time constant τ_E	ms	1.1	1.5	1.3	0.09	0.11	0.2	0.28
Mechanical time constant τ_M	ms	4.7	4.6	3.6	10	10	9.5	9.5
Torque constant K_T	10^{-2} N m A^{-1}	10	17.8	19.3	0.85	0.89	2.3	2.1
Back-e.m.f. constant K_E	10^{-2} V s rad^{-1}	10	17.8	19.3	0.85	0.89	2.3	2.3
Armature resistance R_a	Ω	0.84	1.03	0.47	5.3	3.6	6.3	2.6
Power rate	kW s^{-1}	6.1	11.5	21.5	0.069	0.11	0.29	0.36
Rated continuous torque T	10^{-1} N m	5.9	13	25	0.03	0.05	0.13	0.25
Rated rotational speed Ω	r.p.m.	3000	3000	3000	10500	10500	9200	9200
Rated output P_0	W	185	401	771	3	5	10	20
Rated voltage	V	40.5	68.5	70.9	12	12	24	24

Notes. Original data is converted to SI units.
Rotor. The coil is evenly fixed using epoxy resin and glass tape (see Fig. 2.14). Alnico magnets are used for field system.

$$\text{Power rate} = \frac{\text{Rated torque}}{J^2}.$$

See Section 7.6.1 for power rate.

conversion tables at the end of this book (Appendix) can be used if the reader is unfamiliar with the SI unit system.

Reference for Chapter 2

[1] Lynch, R. (1976). The development of samarium cobalt permanent magnet D.C. servomotors. *IEE Conf. Pub.* **136** (Small Electrical Machines), 5–10.

3. Moving-coil (coreless) motors

As explained in the previous chapter the armature structures of DC motors can be classified into the three fundamental types: slotted, slotless, and moving-coil. Moving-coil motors, which are also known as coreless motors, have recently progressed and are now used in a variety of applications. This chapter will focus on an explanation of the characteristics of moving-coil motors.

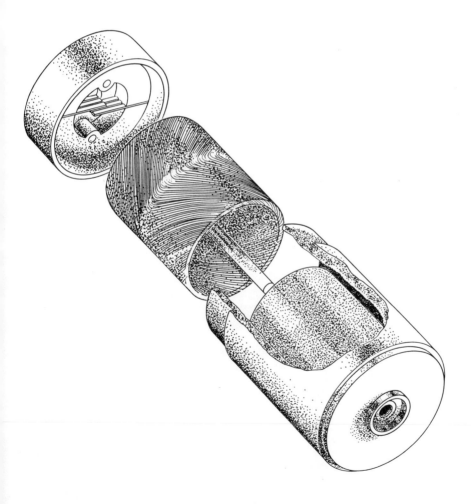

3.1   Classification of moving-coil motors

The moving-coil motor is a DC motor which does not have an iron core in its rotor. The types of moving-coil motors are listed in Table 3.1. The features of these motors will be explained in accordance with the table order.

Table 3.1. Classification of moving-coil motors

Moving-coil motors	Cylindrical types	Outside field type	Faulhaber winding
		Inside field type	Rhombic winding
			Bell winding
			Ball winding
	Disc types	Pancake motors	
		Printed motors	
		Three-coil motors	

This table is re-edited from manufacturers' catalogues.

3.2   Cylindrical outside-field type

This type of motor, which is illustrated in Fig. 3.1, has the smallest mechanical time constant $\tau_M = R_a J / K_T K_E$. In order to obtain a small

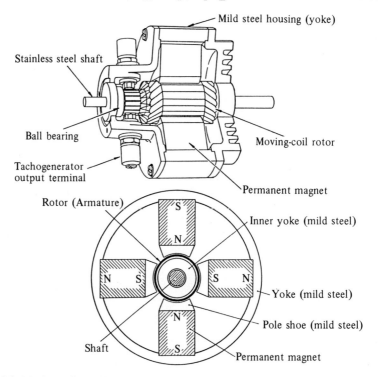

Fig. 3.1. Moving-coil motor which has the field magnets outside the rotor: (above) cutaway view, and (below) cross-sectional view.

value for the mechanical time constant, K_T and K_E must be as large as possible. As explained in Section 1.5, K_T and K_E are essentially the same factor and are proportional to the field flux Φ. More flux is produced from an anisotropic Alnico magnet, which has a high remanence B_r and a low coercive force H_c. Since Alnico magnets are easy to demagnetize, a long magnet, magnetized lengthwise, is used in order to avoid demagnetization. Outside-field motors usually use copper wires in their rotors (see Fig. 2.15), but a smaller value of R_aJ is possible if aluminium wires are used. Some of these motors can have a mechanical time constant of less than 1 ms. The specific resistance and the density of copper and aluminium are shown in Table 3.2. When the adhesive, commutator and shaft elements

Table 3.2. Comparison of specific resistance and density for copper and aluminium

	Copper	Aluminium
Specific resistance ($\mu\Omega$ cm)	1.72	2.75
Density (g cm^{-3})	8.89	2.70

are not taken into account, the τ_M ratio of copper to aluminium is:

$$(\tau_M \text{ ratio}) \quad \text{Cu} : \text{Al} = 1 : 0.486 \qquad (3.1)$$

Figure 3.2 shows a moving-coil motor having a tachogenerator directly coupled to the rotor shaft. A moving-coil armature is used also for the tachogenerator to make the moment of inertia as small as possible.

Machine constants of moving-coil motors   Table 3.3 shows some catalogue data of moving-coil motors produced by two companies.

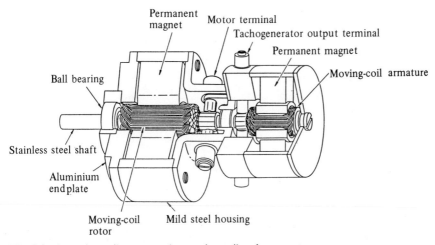

Fig. 3.2. A moving-coil motor and a moving-coil tachogenerator.

Table 3.3. Catalogue data for moving-coil motors

Manufacturers		Sanyo Denki Co., Ltd.			Yasukawa Electric Mfg. Co., Ltd.			
Items		H1008-101	H1009-101	H1420-102	UGSMEM-02A	UGSMEM-02B	UGSMEM-03A	UGSMEM-12B
Inertia J	10^{-6} kg m^2	4.9	3.5	76.5	4	4	3.3	4.65
Electrical time constant τ_E	ms	0.17	0.14	0.3	0.16	0.14	0.15	0.16
Mechanical time constant τ_M	ms	2.3	1.5	2.5	2.0	2.0	1.1	0.75
Torque constant K_T	10^{-2} N m A^{-1}	4.11	4.50	13.3	4.10	8.19	4.49	6.39
Back-e.m.f. constant K_E	10^{-2} V s rad^{-1}	4.11	4.50	13.3	4.10	8.19	4.49	6.39
Armature resistance R_a	Ω	0.7	0.55	0.7	0.80	3.40	0.68	0.67
Power rate	kW s^{-1}	6.5	20	22	11.1	11.1	34.5	28.8
Rated continuous torque T	N m	0.177	0.265	1.32	0.211	0.211	0.28	0.36
Rated speed Ω	r.p.m.	4500	3200	1120	3000	3000	4000	3000
Rated output power P_0	W	85	90	150	43	44	120	114

Cooling the rotor   Since moving-coil motors of the outside-field type are used with a high acceleration, there are two important problems pertaining to the rotor structure:

(1) The coils must sustain a large loading torque, and

(2) Much Joule heat is generated in the rotor because of a large electric current flowing through the coils. There can be a sharp temperature rise since the rotor does not have a heat-absorbing iron core.

To solve these problems, first the coils are reinforced with glass fibre and epoxy resin so that the coils will not easily be deformed by the forces applied. As little glass fibre and epoxy as possible are used in order to prevent an increase in the moment of inertia. Secondly, forced air cooling is employed to effectively remove the generated heat from the wires. The rated current varies by the degree of air cooling. Figure 3.3 shows a motor with ventilation holes.

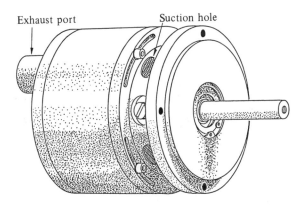

Exhaust port   Suction hole

Fig. 3.3. A moving-coil motor having a suction hole and an exhaust port for air cooling.

3.3   Cylindrical inside-field type

Moving-coil motors of the inner-field type, which are also known as coreless motors, are often used for applications of less than 10 W. But there are micromotors with outputs near 30 W. This type of motor has a magnetic field inside the moving-coil armature. Though the moment of inertia of this rotor is low, the mechanical time constant is not always low, because little magnetic flux is obtained due to the small size of the magnet which must be placed inside the armature. However, coreless motors are extensively used for driving capstans of audio cassette players and video tape recorders, zoom lenses of cameras, etc. because of two outstanding characteristics: (1) very small size and high efficiency, and (2) low cogging.

3.3.1   *Faulhaber winding*

The Faulhaber winding, which is also known as honeycomb winding, was the first type of winding to be employed in the first widely used coreless motor.[1] An exploded view of this type motor is given in Fig. 3.4, and the flux distribution is type shown in Fig. 3.5. Figure 3.6 shows how to wind the rotor and make the coil terminal. This type of winding was invented by F. Faulhaber. Usually this motor uses an Alnico magnet in order to get a high flux. The housing, which also serves as the magnetic path, is made of soft carbon steel. Two bearings are usually placed in the centre hole of the permanent magnet to support the shaft.

For the Faulhaber winding motors, as well as other coreless motors, commutators are made small for the following reasons:

(1) Both commutator and brushes use precious metals, i.e. gold, silver, platinum, and palladium which are resistant to electrochemical processes during operation. Because precious metals are expensive, the size of the brushes must be as small as possible (see Fig. 3.4).

(2) To decrease the peripheral speed of the commutator for stable commutation.

(3) To make the machine size as small as possible.

Mathematical formulae for calculating characteristics of this type of motor are given in reference [1].

3.3.2   *Rhombic winding*

The rhombic winding coreless motor uses rhombic coils. Figure 3.7[2] shows how to wind rhombic coils and make a rotor; Fig. 3.8 shows the finished rotor. Some catalogue data from a manufacturer are given in Table 3.4.

3.3.3   *Bell winding rotor*

Bell winding is a method for making a rotor using rectangular coils, and an example is shown in Fig. 3.9.

3.3.4   *Ball winding rotor*

The ball winding method, shown in Fig. 3.10,[3] makes the rotor like a ball. The permanent magnet producing the field flux is placed inside the ball winding. There is a plastic zone between the magnet and the winding, as illustrated in the cutaway view in Fig. 3.11. The structural details of this type of machine are given in reference [3].

3.4   Disc motors

There are coreless or moving-coil motors where the rotors are flat like a circular plate. There are three main types of disc motors.

Fig. 3.4. Exploded view of a coreless motor employing Faulhaber winding.

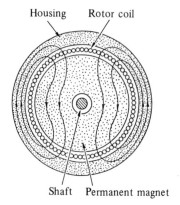

Fig. 3.5. Field flux.

3.4.1 *Pancake motor*

In the pancake motor the winding is usually made of magnetic wires and moulded with resin (as in Fig. 3.12).[4] The type of commutator is identical to the conventional type. Figure 3.13 shows the rotor and the finished motor. One application of these motors can be found in radiator fans.

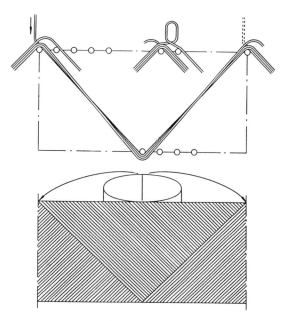

Fig. 3.6. Schematic view of Faulhaber winding and how to make terminals.

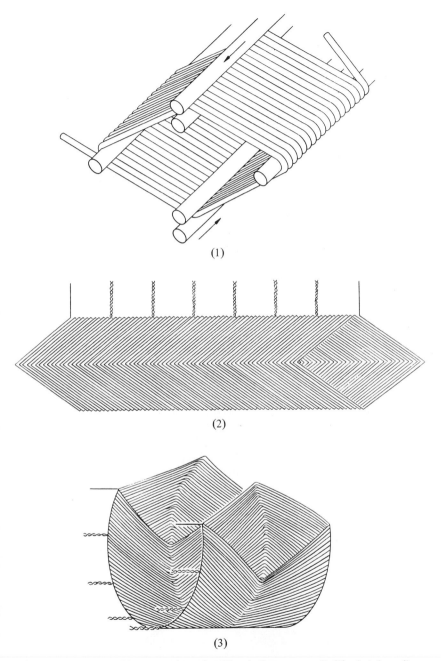

(1)

(2)

(3)

Fig. 3.7. An example of how to make a rhombic winding motor coil; (1) wind the coil on a diamond-shaped jig then move two of the points in opposite directions to each other, (2) make the coil into a flat board shape, and (3) join the ends of the coil together. (After reference [2].)

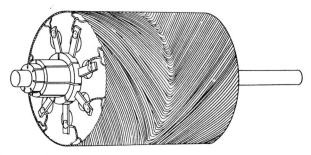

Fig. 3.8. Rhombic winding rotor.

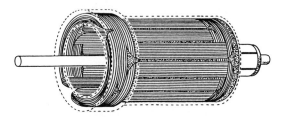

Fig. 3.9. Structure of a bell winding rotor. (The dashed line indicates the resinated part.)

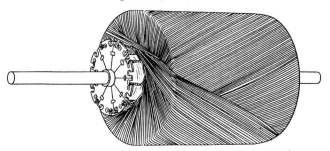

Fig. 3.10. Coreless rotor of the ball winding method.

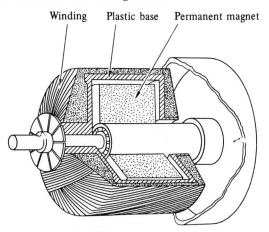

Fig. 3.11. Cutaway view of a ball winding rotor.

Table 3.4. Catalogue data for rhombic winding coreless motors

Items		LN12 -K91N1	LN20 -N1N1	LN22 -M11N1	LN40 -J21N1	LN30 -H21N1	LN30 -J31N1
	Manufacturer			Canon Seiki Co., Ltd.			
Inertia J	10^{-6} kg m^2	0.011	0.15	0.31	8.0	2.2	2.7
Mechanical time constant τ_M	ms	27	21	15	56	43	38
Torque constant K_T	10^{-2} N m A^{-1}	0.18	1.10	1.20	6.62	1.12	1.76
Back-e.m.f. constant K_E	10^{-2} V s rad	0.18	1.10	1.20	6.62	1.12	1.76
Armature resistance R_a	Ω	8.1	17.2	7.4	13.5	2.2	4.7
Rated voltage	V	1.5	6	6	24	6	12
Rated continuous torque T	10^{-1} N m	9.8×10^{-4}	0.015	0.020	0.147	0.029	0.049
Rated rotational speed Ω	r.p.m.	4600	2900	3700	1880	4350	5700
Rated output P_0	W	0.47	4.6	7.7	3.0	13	29
Weight	g	10	50	70	290	120	135

Note. Inside-field type.
This table is re-edited from the manufacturer's catalogue.

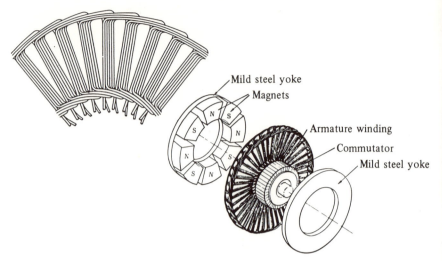

Fig. 3.12. Coils, field magnets, and commutators of a pancake motor. (After reference [4], reproduced by permission of the Institution of Electrical Engineers, and by courtesy of Mr P. Campbell.)

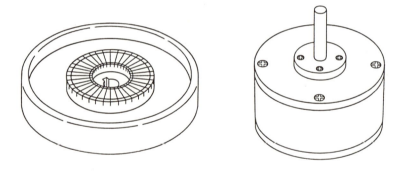

Fig. 3.13. Rotor and a finished pancake motor.

Fig. 3.14. Printed winding.

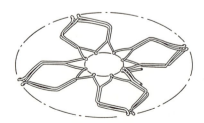

Fig. 3.15. Coil arrangement for a printed rotor.

(a)

(b)

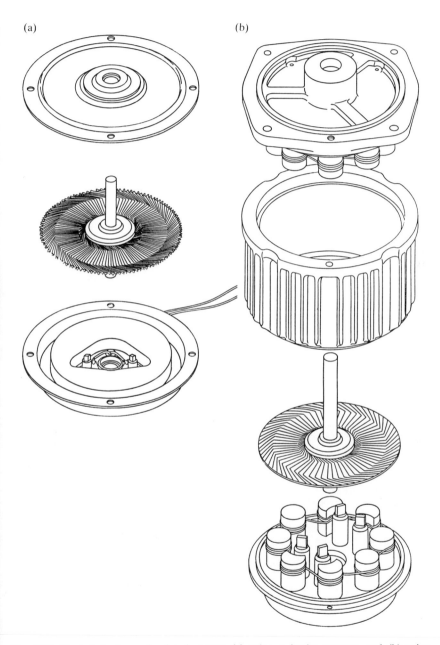

Fig. 3.16. Exploded views of printed motors (a) using a ferrite magnet, and (b) using Alnico magnets.

3.4.2 *Printed motor*

The coil for this armature is shown in Fig. 3.14. The coils are stamped from pieces of sheet copper and then welded, forming a wave winding. Figure 3.15 shows how the wave winding is made. When this motor was invented by J. Henry Baudot, the armature was made using a similar method to that by which printed boards are manufactured. Hence this is called the printed motor.

The field flux of the printed motor can be produced using either Alnico or ferrite magnets. Figure 3.16 shows a disassembled diagram of each style of motor.

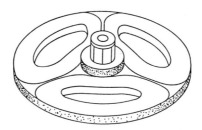

Fig. 3.17. Three-coil type disc motor.

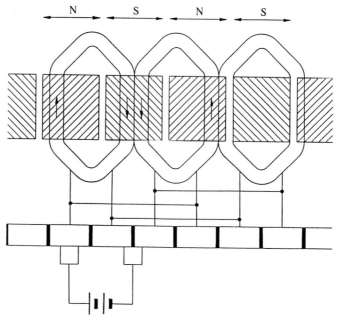

Fig. 3.18. Connection and current distribution in three-coil disc motor.

3.4.3 Three-coil motor

Figure 3.17 shows a micromotor with a special structure and the following characteristics:

(1) There are four magnetic poles in the stator, whilst the armature has a three-coil arrangement.

(2) This winding arrangement is similar to the two-pole type shown in Fig. 4.12 (see p. 68). However, the coil connection shown in Fig. 3.18 differs from the ordinary lap winding or wave winding.

3.5 Characteristics of moving-coil motors

It is useful to understand the differences between moving-coil and slotted DC motors so that their characteristics can be positively utilized in proper applications. One major difference between the slotted and moving-coil motors is that the rotor of the latter has no iron core. This means that a moving-coil motor has two features: (1) the rotor is light, and (2) cogging is very low because there are no slots or teeth on the core.

3.5.1 The meaning of light weight

It is said that a light rotor has a good dynamic response. This is partly, but not always true. Moreover, even if a motor exerts a high response with no load, the same motor will not always respond well in its final application. This will be explained as follows.

The printed motor does not have a low moment of inertia The mechanical time constant $(\tau_M = JR_a/K_T K_E)$ is the index which indicates good response to a change in the speed command. The moment of inertia J is one of the factors affecting this constant, and it is desirable that J is as small as possible. Since the moment of inertia is proportional to the fourth power of the radius, the moment of a printed motor or pancake motor is large. When three different types of rotor with identical output power are compared, the inertia of the printed motor is close to that of the slotted rotor (see Table 3.5). Therefore, to obtain high response, it is desirable to use a cylindrical rotor instead of a disc type.

A low inertia motor does not always mean a low time constant Table 3.6 shows the constants for various types of moving-coil DC motors.

Table 3.5. Comparison of inertias of three different typical rotors which offer similar values of output power

	Printed rotor	Cylindrical moving-coil rotor	Slotted rotor
Inertia (10^{-6} kg m^2)	160	3.8	235
Rated output power (W)	200	198	200

Table 3.6. Comparison of the catalogue data for various types of moving-coil motors

Manufacturers and models	Cylindrical rotor, outside field — Tamagawa Seiki Co., Ltd.				Cylindrical rotor, inside field — Namiki Precision Jewel Co., Ltd.				Printed rotor — Yasukawa Electric Mfg. Co., Ltd.	
	M-1600 -A	M-1020 -A	TS3516 -E17	TS3513 -E5	12-2006	12-3006	16-2004	16-2501	UGPMEN -90DAB	UGPMEE -09B12
Inertia J (10^{-6} kg m^2)	2.2	3.3	28	2.75	0.0039	0.016	0.022	0.043	43	34
Electrical time constant τ_E (ms)	0.1	0.12	0.4	0.12	0.008	0.005	0.007	0.012	0.043	0.06
Mechanical time constant τ_M (ms)	0.5	3.8	1.3	3.2	3.5	5.8	13.9	6.6	8.0	37
Torque constant K_T (10^{-2} N m A^{-1})	6.7	2.54	16.4	5.41	0.69	1.12	0.29	2.02	5.0	3.2
Back-e.m.f. constant K_E (10^{-2} V s rad^{-1})	6.7	2.54	16.4	5.41	0.69	1.12	0.29	2.02	5.0	3.2
Armature resistance R_a (Ω)	0.8	0.7	1.23	3.2	39	46	6	63	0.46	1.02
Power rate (kw s^{-1})	280	12.3	40.8	5.9	—	—	—	—	1.3	0.15
Rated continuous torque T (10^{-1} N m)	8.5	3.3	10.8	1.27	S0.0012	S0.028	S0.020	S0.038	2.38	0.72
Rated rotational speed Ω (r.p.m.)	4500	4000	2100	3850	N10000	N10000	N12600	N5600	4000	4000
Rated output P_0 (W)	375	130	200	50	0.23	0.53	0.48	0.4	100	30
									Alnico magent	Ferrite magnet

S: Starting torque
N: No-load torque
Rotor: Faulhaber winding

This table is re-edited from manufacturers' catalogues.

Regarding the mechanical time constant we can notice the following:
(1) For the cylindrical rotor

τ_M is especially small in the outside-field type motors.
τ_M is large in the inside field type motors. This is especially true with low-output motors.

(2) For the printed rotor

τ_M is relatively small when an Alnico magnet is used.
τ_M is large when a ferrite magnet is used.

Thus we can conclude that the outside-field type cylindrical motor is the only motor with a small mechanical time constant.

3.5.2 *The meaning of no slots and no teeth*

The moving-coil motor has two merits due to the absence of slots and teeth as explained in the following:

(1) *The clogging is small* In the slotted motor, since the slots and teeth are alternately passed by the field magnet, cogging is generated due to reluctance variation during rotation. In a moving-coil motor, the reluctance change is non-existent since there is no slotted core in the rotor. Therefore, a moving-coil motor provides a smooth, coggingless rotation.

(2) *Inductance is small and commutation is good* Armature inductance in a slotted rotor is a sum of the following: (i) the self-inductance of the conductor in the slots (see Fig. 3.19), (ii) the self-inductance caused by the flux leakage to the field magnet zone, and (iii) the self-inductance caused by the leakage flux at the edge of the armature coils. Of these factors, the first gives rise to the major part of the total self-inductance in the slotted rotor. Because in the moving-coil motor the first factor is small, the inductance is very small as compared to the slotted motor. This means that even if metallic brushes are used for a coreless motor, commutation is carried out fairly efficiently, i.e. few sparks are generated. For a low voltage (battery driven) DC motor, metallic brushes are required to make a small potential drop across the brushes. The combination of a large self-inductance in a slotted rotor and small potential drop across the brushes generates large sparks.

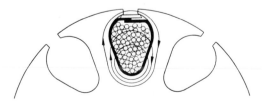

Fig. 3.19. Leakage flux in slots.

References for Chapter 3

[1] Weissmantel, H. (1976). Einige Grundlagen zur Berechnung bei der Anwendung schnell hochlaufender, trägheitsamer Gleichstromkleinstmotoren mit Glockenläufer. *Feinwerktechnik & Messtechnik* **85,** 165–74. Carl Hanser Verlag.

[2] Heyraud, M. (1977). Zuverlässigkeit und Lebensdauer von Gleichstrommotoren: das LEE-System. *Ibid.* **85,** 121–3.

[3] Hofmeester, J. H. M. and Koutstaal, J. P. (1973). Moving coil motors. *Philips Tech. Rev.* **33,** 244–8.

[4] Campbell, P. (1974). Principles of permanent magnet axial-field D.C. machines. *Proc. IEE* **121,** (12), 1489–94.

4. Principles and fundamental structures of brushless DC motors

Conventional DC motors are highly efficient and their characteristics make them suitable for use as servomotors. However, their only drawback is that they need a commutator and brushes which are subject to wear and require maintenance. When the functions of commutator and brushes were implemented by solid-state switches, maintenance-free motors were realized. These motors are now known as brushless DC motors. This chapter will focus on an explanation of the fundamental principles and some practical structures of brushless DC motors.

4.1 Basic brushless DC motor

In conventional DC motors, the armature is the rotor, and the field magnets are placed in the stator. A brushless DC motor of this structure is very difficult to make. The construction of modern brushless DC motors is very similar to the AC motor, known as the permanent magnet synchronous motor (see Fig. 4.1). The armature windings are part of the stator, and the rotor is composed of one or more magnets. The windings in a brushless DC motor are similar to those in a polyphse AC motor, and the most orthodox and efficient motor has a set of three-phase windings and is operated in bipolar excitation (see Fig. 4.2). Brushless DC motors are different from AC synchronous motors in that the former incorporates some means to detect the rotor position (or magnetic poles) to produce signals to control the electronic switches. The most common position/pole sensor is the Hall element, but some motors use optical sensors.

By examining a simple three-phase unipolar-operated motor, one can easily understand the basic principles of brushless DC motors. Figure 4.2 illustrates a motor of this type that uses optical sensors (phototransistors) as position detectors. Three phototransistors PT1, PT2, and PT3 are placed on the end-plate at 120° intervals, and are exposed to light in sequence through a revolving shutter coupled to the motor shaft.

As shown in Fig. 4.2, the south pole of the rotor now faces the salient pole P2 of the stator, and the phototransistor PT1 detects the light and turns transistor Tr1 on. In this state, the south pole which is created at the

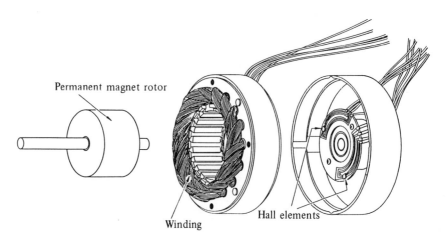

Permanent magnet rotor

Winding

Hall elements

Fig. 4.1. Disassembled view of a brushless DC motor: permanent magnet rotor, winding, and Hall element.

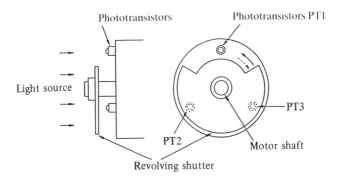

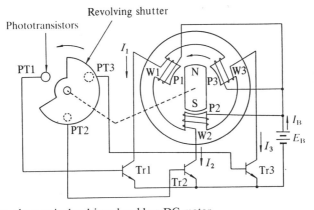

Fig. 4.2. Three-phase unipolar-driven brushless DC motor.

salient pole P1 by the electrical current flowing through the winding W1 is attracting the north pole of the rotor to move it in the direction of the arrow (CW). When the south pole comes to the position to face the salient pole P1, the shutter, which is coupled to the rotor shaft, will shade PT1, and PT2 will be exposed to the light and a current will flow through the transistor Tr2. When a current flows through the winding W2, and creates a south pole on salient pole P2, then the north pole in the rotor will revolve in the direction of the arrow and face the salient pole P2. At this moment, the shutter shades PT2, and the phototransistor PT3 is exposed to the light. These actions steer the current from the winding W2 to W3. Thus salient pole P2 is de-energized, while the salient pole P3 is energized and creates the south pole. Hence the north pole on the rotor further travels from P2 to P3 without stopping. By repeating such a switching action in the sequence given in Fig. 4.3, the permanent magnet rotor revolves continuously.

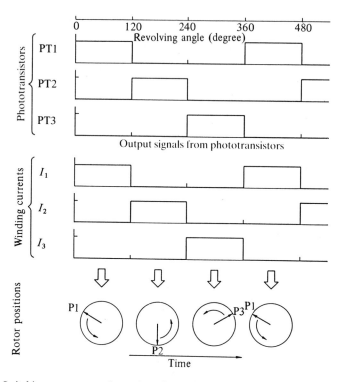

Fig. 4.3. Switching sequence and rotation of stator's magnetic field.

4.1.1   *Method of reversing the direction of rotation*

In order to reverse the direction of rotation of a conventional DC motor, the terminal voltage simply has to be reversed. However, such an action will not reverse a brushless DC motor which uses semiconductor devices like transistors, because most semiconductor devices are unidirectional switches. Therefore, some circuitry means is necessary when the motor is to be driven in either direction.

In Fig. 4.3, the connections between the phototransistors (PT1, PT2, and PT3) and the transistors (Tr1, Tr2, and Tr3) are arranged as

> PT1—Tr1 for controlling current through W1
> PT2—Tr2 for controlling current through W2
> PT3—Tr3 for controlling current through W3.

These connections make the motor rotate counter-clockwise. If the connections are changed over to

> PT1—Tr3
> PT2—Tr1
> PT3—Tr2

the rotational direction will be reversed.

Table 4.1. Switching sequences for CW and CCW directions

Revolving direction Switching sequence		CCW				CW			
		1	2	3	4	1	2	3	4
Phototransistors	PT1	1	0	0	1	1	0	0	1
	PT2	0	1	0	0	0	0	1	0
	PT3	0	0	1	0	0	1	0	0
Transistors	Tr1	1	0	0	1	0	0	1	0
	Tr2	0	1	0	0	0	1	0	0
	Tr3	0	0	1	0	1	0	0	1

The change-over between those sets of connections can be implemented by a logic-gate circuit. These sequential relationships are given in Table 4.1.

4.2 Three-phase bipolar-driven motor

When a three-phase (brushless) motor is driven by a three-phase bridge circuit, the efficiency, which is the ratio of the mechanical output power to the electrical input power, is the highest, since in this drive an alternating current flows through each winding as in an AC motor. This drive is often referred to as 'bipolar drive'. Here, 'bipolar' means that a winding is alternatively energized in the south and north poles.

We shall now survey the principle of the three-phase bridge circuit of Fig. 4.4. Here too, we use the optical method for detecting the rotor

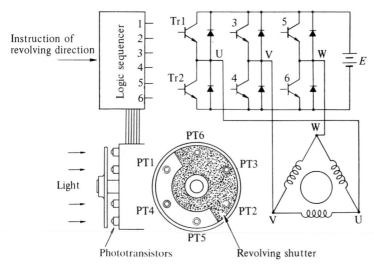

Fig. 4.4. Three-phase bipolar-driven brushless motor.

position; six phototransistors are placed on the end-plate at equal intervals. Since a shutter is coupled to the shaft, these photo elements are
exposed in sequence to the light emitted from a lamp placed in the left of
the figure. Now the problem is the relation between the ON/OFF state of
the transistors and the light detecting phototransistors. The simplest
relation is set when the logic sequencer is arranged in such a way that
when a phototransistor marked with a certain number is exposed to light,
the transistor of the same number turns ON. Figure 4.4 shows that
electrical currents flow through Tr1, Tr4, and Tr5, and terminals U and
W have the battery voltage, while terminal V has zero potential. In this
state, a current will flow from terminal U to V, and another current from
W to V as illustrated in Fig. 4.5. We may assume that the solid arrows in
this figure indicate the directions of the magnetic fields generated by the
currents in each phase. The fat arrow in the centre is the resultant
magnetic field in the stator.

The rotor is placed in such a position that the field flux will have a 90°
angle with respect to the stator's magnetic field as shown in Fig. 4.5. In
such a state a clockwise torque will be produced on the rotor. After it
revolves through about 30°, PT5 is turned OFF and PT6 ON which
makes the stator's magnetic pole revolve 60° clockwise. Thus when the
rotor's south pole gets near, the stator's south pole goes away further to
create a continuous clockwise rotation. The ON–OFF sequence and the
rotation of the transistor are shown in Fig. 4.6.

The rotational direction may be reversed by arranging the logic sequencer in such a way that when a photodetector marked with a certain
number is exposed to light, the transistor of the same number is turned
OFF. On the other hand, when a photodetector is not exposed to light,
the transistor of the same number is turned ON.

In the positional state of Fig. 4.4, Tr2, 3, and 6 are ON, and the battery
voltage E appears at terminal V, while U and W have zero electric
potential. Then, as shown in Fig. 4.7(a), the magnetic field in the stator is
reversed, and the rotor's torque is counter-clockwise. After the motor
revolves about 30°, Tr2 turns OFF and Tr1 ON. At this point, the field
has revolved 60° and becomes as shown in (b). As the rotor produces

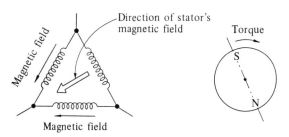

Fig. 4.5. Stator's magnetic field in the shutter state of Fig. 4.4, and the direction of torque.

ON-OFF sequence	1	2	3	4	5	6
Tr 1	1	1	1	0	0.	0
2	0	0	0	1	1	1
3	0	0	1	1	1	0
4	1	1	0	0	0	1
5	1	0	0	0	1	1
6	0	1	1	1	0	0

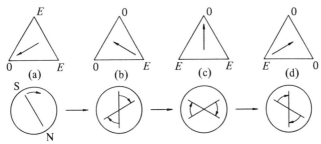

Fig. 4.6. Clockwise revolutions of the stator's magnetic field and rotor.

ON-OFF sequence	1	2	3	4	5	6
Tr 1	0	1	1	1	0	0
2	1	0	0	0	1	1
3	1	1	0	0	0	1
4	0	0	1	1	1	0
5	0	0	0	1	1	1
6	1	1	1	0	0	0

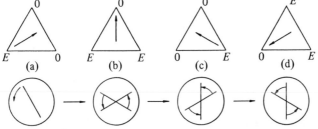

Fig. 4.7. Counter-clockwise revolutions of the stator's magnetic field and rotor.

another counter-clockwise torque, the counter-clockwise motion continues and the field becomes as shown in (c). This action is repeated in the sequence of (a) → (b) → (c) → (d)...to produce a continuous counter-clockwise motion.

4.3 Comparison of conventional and brushless DC motors

Although it is said that brushless DC motors and conventional DC motors are similar in their static characteristics, they actually have remarkable differences in some aspects. When we compare both motors in terms of present-day technology, a discussion of their differences rather than their similarities can be more helpful in understanding their proper applications. Table 4.2 compares the advantages and disadvantages of these two types of motors. When we discuss the functions of electrical motors, we should not forget the significance of windings and commutation. Commutation refers to the process which converts the input direct current to an alternating current and properly distributes it to each winding in the armature. In a conventional DC motor, commutation is undertaken by brushes and commutator; in contrast, in a brushless DC motor it is done by using semiconductor devices such as transistors.

Table 4.2. Comparison of conventional and brushless DC motors

	Conventional motors	Brushless motors
Mechanical structure	Field magnets on the stator	Field magnets on the rotor Similar to AC synchronous motor
Distinctive features	Quick response and excellent controlability	Long-lasting Easy maintenance (usually no maintenance required)
Winding connections	Ring connection The simplest: Δ connection	The highest grade: Δ or Y-connected three-phase connection Normal: Y-connected three-phase winding with grounded neutral point, or four-phase connection The simplest: Two-phase connection
Commutation method	Mechanical contact between brushes and commutator	Electronic switching using transistors
Detecting method of rotor's position	Automatically detected by brushes	Hall element, optical encoder, etc.
Reversing method	By a reverse of terminal voltage	Rearranging logic sequencer

4.3.1 *Difference in commutation*

In Fig. 4.8, to explain the function of a brush in a DC motor, it is replaced with two switches or transistors. Consider the commutator

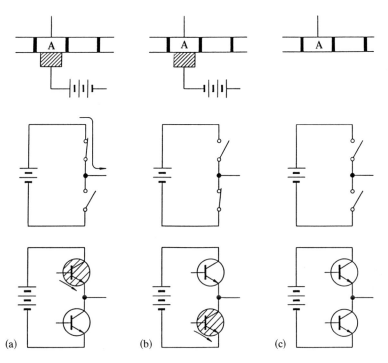

Fig. 4.8. DC motor's brushes replaced with mechanical switches or transistors: (a) commutator segment A is touching the positive brush, (b) touching the negative brush, (c) touching neither.

segment A which is in either of the three states: (a) touching the positive brush, (b) touching the negative brush, or (c) touching neither. As seen in Fig. 4.8, one commutator segment corresponds to two transistors.

Attention should be paid to the action of the coil inductance in the conventional DC motor. In Fig. 4.9(a), switch S1 is closed and the current is supplied to the coil. In Fig. 4.9(b) the current is apparently cut off when S1 is opened, but because of inductance, a high voltage emerges across the air gap in the switch and sparking occurs; the current continues to flow through the air gap for a short time. Returning to the use of brushes as shown in Fig. 4.9(c), we note that a spark can occur when the brush separates from the commutator segment A. Weak sparks will not produce serious damage to the commutator segment or the brush. However, if sparks are strong and repeated during the operation, the brushes and commutator will be damaged.

Generally, when current (i) flows into a circuit whose inductance is L, electromagnetic energy ($1/2Li^2$) will be stored in it. When the current is

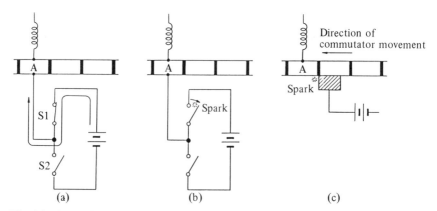

Fig. 4.9. Generation of sparks between a commutator segment and a brush.

cut off, this energy is discharged as a spark. The greater the energy, the larger the spark.

A smaller inductance will reduce sparking. That is, the number of turns of the coil connected to one commutator segment should be small for sparkless commutation. In conventional motors, the windings are separated into many coils, and in some motors, there are more coils than there are commutator segments employing the construction shown in Fig. 4.10.

Figure 4.11 shows a commutator circuit which uses two transistors. In Fig. 4.11(a), Tr1 is ON and a current is being supplied from the power supply through this transistor to the coil L. As shown in (b), when Tr1 turns OFF, the current will flow through diode D2 for a transient period until it falls to zero. If this diode is not provided, the current will decrease all at once. Hence a high voltage, which is $L(di/dt)$, will be produced in the coil and will be applied across the collector and emitter of Tr1, and

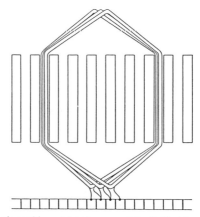

Fig. 4.10. For reduction of sparking at brushes a coil is divided into several subcoils.

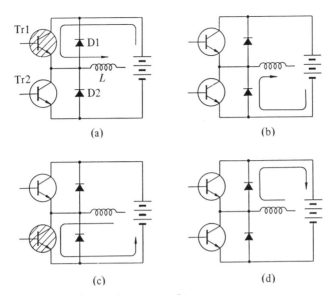

Fig. 4.11. Commutator circuit using two transistors.

will damage the transistor. Thus, in a brushless DC motor the electromagnetic energy stored in the coil can return to the power supply through this diode. Therefore, there is no spark. We see that diode D2 is needed to protect Tr1 from being damaged, and D1 protects Tr2.

4.3.2 *Difference in winding*

In Fig. 4.8, one commutator segment is equivalent to a combination of two transistors and two diodes. Therefore, if a conventional DC motor having many commutator segments is to be converted to a brushless DC motor based on this equivalency principle, the cost of such electronic devices would be too high. For economic reasons it is desirable that the smallest number of semiconductor devices be used to drive a brushless DC motor. A conventional DC motor must have at least three commutator segments so that its rotational direction is determined by the voltage polarity applied. Figure 4.12(a) shows this type of armature. A motor which has such an armature is known as a three-segment motor. The cross-sectional diagram in Fig. 4.12(b) is that of a three-phase brushless DC motor, which has three slots and the same number of teeth. Since each phase winding is wound around one tooth, this type of winding is called the 'concentrated winding'. The motor construction is simple when this type of winding is employed, but motion will not be smooth. To make the rotation smooth, we must employ a coil arrangement known as a distributed winding. For such a winding, the number of slots will be increased. Figure 4.13 shows a four-pole distributed winding which uses

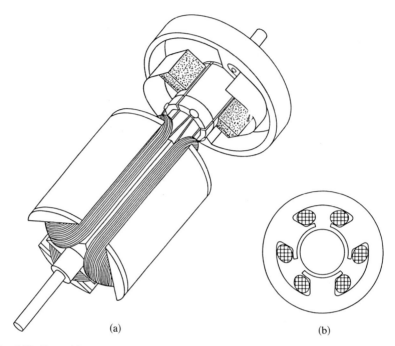

Fig. 4.12. From (a) a conventional motor to (b) a brushless motor converted from (a).

24 slots. Since the distributed winding has long been employed in squirrel induction motors, brushless DC motors are similar to AC motors in terms of machine structure.

Why are three-phase distributed windings not used in conventional DC motors? The answers are:

(1) Three-phase windings increase the coil inductance and cause sparks. The three-slot motor shown in Fig. 4.12(a), which is the simplest

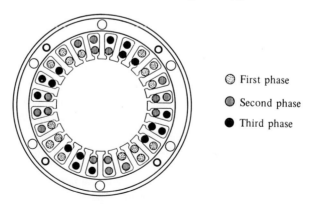

⊛ First phase

⦿ Second phase

● Third phase

Fig. 4.13. Coil arrangement employing three-phase four-pole distributed winding.

three-phase motor, produces a large cogging torque. Therefore when high performance is required, use of such a motor is not recommended.

(2) Many slots are required to decrease the cogging torque and lessen the adverse effects of sparks. Moreover, to minimize sparks, coils are divided into several sub-coils, each being joined at a different segment as shown in Fig. 4.10.

4.4 Detection of rotor position and the use of Hall elements

A brushless DC motor incorporates some means of detecting the pole/position on its rotor. The position sensors used nowadays are:
  (a) Hall elements,
  (b) Light-emitting diodes and phototransistors (or photodiode),
  (c) Inductors sensitive to inductance variation.
Of these, (a) is used extensively followed by (b). Category (c) was once used, but recently its use has been declining. The principles of the Hall element and Hall ICs will now be discussed.

4.4.1 Hall elements

In Fig. 4.14, when an electrical current I_c flows downwards in a semiconductor pellet which is placed in a magnetic field perpendicular to the pellet surface, an electromotive force V_H is created in the pellet in a direction perpendicular to both current I_c and magnetic induction B. Since the electromagnetic force acts on charged particles (electrons or holes) according to Fleming's left-hand rule, the charged particles are biased to the left side of the semiconductor pellet. When it is a semiconductor pellet, the polarity of the electromotive force depends on whether the semiconductor is the p-type or n-type. The magnitude of the electromotive force V_H, which is called the Hall voltage, is given by the following equation:

$$V_H = \frac{1}{d} B I_c R_H,$$ (4.1)

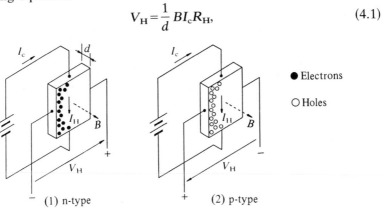

(1) n-type   (2) p-type

● Electrons
○ Holes

Fig. 4.14. Hall effect.

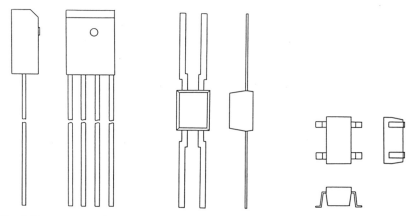

Fig. 4.15. Typical Hall elements.

where R_H is the Hall constant ($m^3\,C^{-1}$),
I_c is the electrical current (A),
B is the flux density (magnetic induction) (T), and
d is the thickness of the semiconductor pellet (m).

This phenomenon was discovered by E. H. Hall in 1878 from an experiment using a metal segment, and is called the Hall effect. The Hall effect is strong in some specific metal compounds or semiconductors. Semiconductor devices which are made for use in detecting magnetic fields are called Hall elements or Hall generators. In modern brushless motors, n-type InSb (indium–antimony) is extensively used, as well as GaAs (gallium–arsenide). Figure 4.15 shows typical Hall elements available on the market, and Fig. 4.16 shows typical sizes of Hall elements.

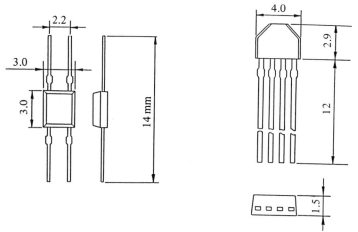

Fig. 4.16. Typical sizes of Hall elements (mm).

4.4.2 Principle of position detection using Hall elements

Figure 4.17 shows an equivalent circuit for a Hall element expressed as a four-terminal network. When a current I_c, which is called the control current or bias current, flows from terminal 3 to 4 in the Hall element exposed to a magnetic field which is perpendicular to the element plane, a voltage V_H is generated across terminals 1 and 2 as explained before. When terminal 4 is taken as the reference point, the potentials at terminals 1 and 2 are $V_H/2$ and $-V_H/2$, respectively, where $R_1 = R_2$ and $R_3 = R_4$ are assumed. Moreover, the polarity reverses as the flux direction reverses. Figure 4.18 shows these relationships. Thus when a Hall element is placed near a permanent magnet rotor, the Hall element can accurately detect the pole positions and the flux density, providing output voltages V_{H1} and V_{H2}.

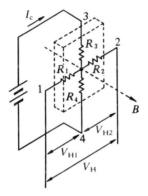

Fig. 4.17. Equivalent circuit for a Hall element.

4.4.3 Practical methods of position detection

Figure 4.19(a) shows the simplest brushless DC motor using one Hall element placed as shown in Fig. 4.19(b). The output signals from the Hall element operate two transistors Tr1 and Tr2 to control the electrical currents in stator windings W1 and W2. Figure 4.20(a)–(c) shows the state of the rotor as it revolves, which is explained as follows:

(a) The Hall element detects the north pole of the rotor magnet, and winding W2 is energized to produce the south pole which drives the rotor in the CCW direction.

(b) Since no magnetic field is applied to the Hall element in this positional relation, both transistors are in the OFF state, and no currents flow in W1 or W2. The rotor continues to revolve due to inertia.

(c) The Hall element detects the south pole of the rotor, and winding W1 is energized to create the south pole which attracts the north pole of the rotor to continue the CCW motion.

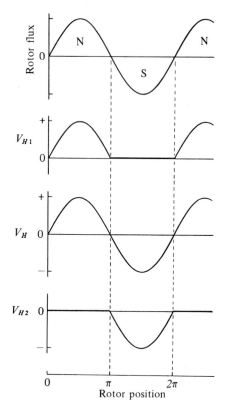

Fig. 4.18. Output waveforms from a Hall element.

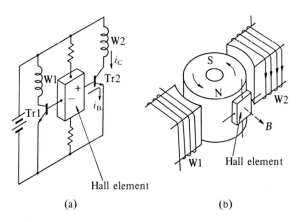

(a)   (b)

Fig. 4.19. Basic principles of the brushless DC motor using a Hall element.

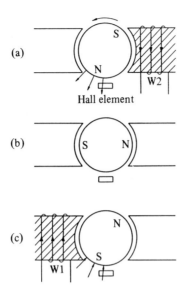

Fig. 4.20. Torque generation, revolution, and switching.

4.4.4 Hall ICs

As explained before, Hall elements must be accompanied by one or more transistors to amplify the output signals. Recently a Hall element and some electronic circuits were formed and moulded on the same chip. This is the Hall IC. A typical appearance of the Hall IC, and a block diagram of its function are shown in Figs. 4.21(a) and (b). The output signal from the Hall element is amplified once by an operational amplifier and its output signal is further processed in the output stage. The output signal from the Hall IC is used to drive a power transistor to control the winding currents. There are two types of Hall IC, namely the linear type and the switching type. Figure 4.22 shows the sensibility characteristics of these two types. Which type is selected depends on the motor structure and application.

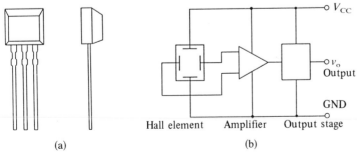

Fig. 4.21. A Hall IC and its function: (a) external appearance, (b) block diagram of Hall IC.

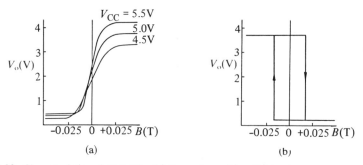

Fig. 4.22. Characteristics of Hall ICs: (a) linear type, (b) switching type.

Figure 4.23 shows an example of an outer-rotor motor which uses three Hall ICs. The stator windings are of a three-phase type and driven in the unipolar energization. The Hall ICs are set up in this motor to detect the main flux from the rotor, and the switching sequence is the same as explained in Section 4.1 using opto-electronic means for rotor position detection.

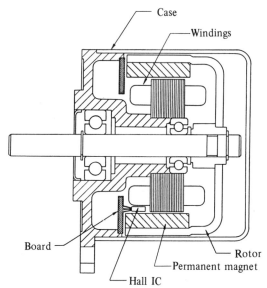

Fig. 4.23. Outer-rotor brushless DC motor.

4.5   Elimination of deadpoints in two-phase motors

Section 4.4.3 explained the principle of a two-phase motor which uses one Hall element. This type of motor, however, has the following two drawbacks.

(1) There are two deadpoints at which the Hall element does not

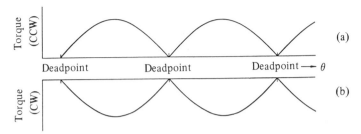

Fig. 4.24. Relationship between torque and revolving angle in a plain two-phase motor.

experience the magnetic flux, and as a result no current flows in the windings to produce a torque. Hence, when the motor carries a frictional load, there is a possibility that it will stop at a deadpoint and be unable to start again. When the frictional load is small, the rotor may be able to pass through the deadpoint due to inertia (see Fig. 4.24).

(2) Since the back e.m.f. is small at the low torque positions, a large current will flow and increase the conduction loss. Hence the motor efficiency, which is the ratio of the output mechanical power to the input electrical power, is not high.

For the motors which are used in practical applications these problems have been cleared up. The type of motor which is most expensive but has the highest efficiency is a three-phase type driven in the bipolar energization scheme as explained in Section 4.2. On the other hand, the two-phase motor having no deadpoints is the cheapest brushless DC motor. Between these two extremes there are several classes of brushless DC motors. As for the elimination of the deadpoints there are two major methods: one uses a polyphase structure, and the other utilizes a space harmonic magnetic field.

4.5.1 *Polyphase motors*

Figure 4.25 shows an example of a three-phase motor. When a direct current is given to each phase (W1, W2, and W3) of the windings, and the rotor is rotated by an external force, the torque which appears at the shaft, due to the interaction between the winding current and the magnet flux, will be as shown in Fig. 4.26. This is close to a sinusoidal wave, each phase being shifted relative to one another by 120°. If the Hall elements are used to steer the currents in the windings in sequence to energize each phase at the largest torque position, the darkened curve in Fig. 4.26 will be obtained. This example is a three-phase motor in which each phase carries a current in one direction for a 120° period in sequence. Hence this motor is called a three-phase unipolar motor. The type shown in Section 4.2 employs a bipolar scheme in which the current can flow in either direction. There is also a unipolar four-phase motor.

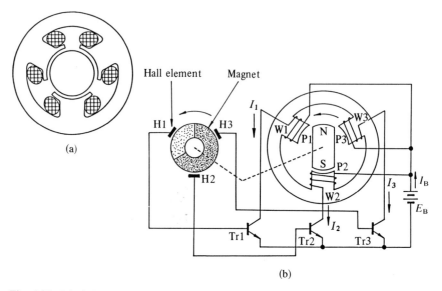

Fig. 4.25. Principles of a three-phase unipolar motor: (a) cross-section, (b) principles of operation.

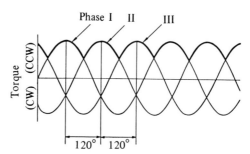

Fig. 4.26. Relationship between torque and revolving angle when a direct current flows in each phase.

4.5.2 *Space-harmonic type*

Consider the motor of Fig. 4.27 which is an improved type of the plain two-phase, two-pole motor shown in Figs. 4.19 and 4.20. An additional four-pole magnet is coupled to the rotor, and the stator also has a four-pole magnet in a positional relationship as shown in the figure. Figure 4.28 shows the relationship between the torque and revolving angle for this motor. The two-pole torque curve, which is indicated by A, is identical to Fig. 4.24 and has two deadpoints. Curve B is the space-harmonic torque curve created by the newly installed permanent magnets. The two-pole torque is always counter-clockwise because of the commutation by the Hall element. The four-pole torque varies alternately in both

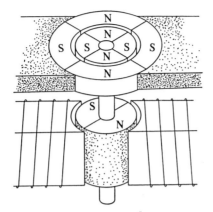

Fig. 4.27. A two-phase motor using a space-harmonic torque.

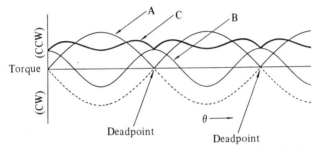

Fig. 4.28. Elimination of deadpoints by superposing a four-poled harmonic torque on the two-poled fundamental torque.

directions (CW and CCW). Since the overall torque, which is shown by curve C, is the sum of both torques, it does not have any deadpoints.

This type of motor in which two additional magnets are set up has not been used in a practical application. The methods of exploiting this effect are as follows:

(1) The method which has auxiliary salient poles in its stator and an auxiliary magnet in the rotor, the magnet having twice as many magnetic poles as the main magnet.

(2) The method which uses a second harmonic magnetization superimposed on the fundamental-wave magnetization in the rotor, the winding pitch being moved from 180° on the stator.

(3) The method which uses a non-uniform air gap to create a second harmonic torque.

5. Modern brushless DC motors used in information instruments

In the previous chapter, we studied the principles of brushless DC motors and the differences between brushless and conventional DC motors. We also looked at some of the fundamental principles of brushless DC motors which use Hall elements as pole sensors. In this chapter we shall examine some motors which are in use in information instruments.

5.1 Three-phase bipolar-driven motors

The basic description of the brushless DC motor in Section 4.2 was for a three-phase bipolar-driven motor (see Fig. 4.4). The windings employed in this type of motor are the same as those used in ordinary three-phase AC motors. Since alternating currents flow in the windings, the efficiency of this brushless DC motor is high.

The motor discussed in Section 4.2 has delta-connected windings, but it may also have Y-connected windings. Figure 5.1(a) shows a practical circuit which is used in a laser-beam printer or a hard-disc drive. As shown in Fig. 5.1(b), three Hall elements are placed at intervals of 60° for detection of the rotor's magnetic poles. Because this motor has four magnetic poles, a mechanical angle of 60° corresponds to an electrical angle of 120°. Figure 5.2 shows voltage waveforms at various parts of the circuit in Fig. 5.1.

Here an application of this type of motor to the laser-beam printer is taken. In a laser printer, a polygon mirror is coupled directly to the motor shaft and its speed is controlled very accurately in the range from 5000 to 40 000 r.p.m. When an intensity-modulated laser beam strikes the revolving polygon mirror, the reflected beam travels in different directions according to the position of the rotor at that moment. Therefore, this

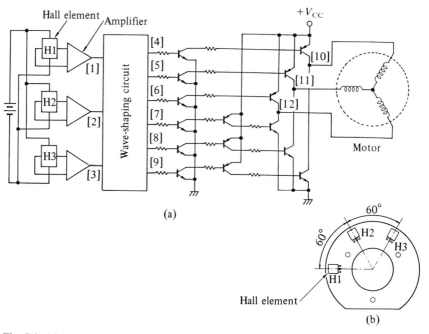

Fig. 5.1. (a) Practical circuit for a three-phase bipolar-driven motor, and arrangement of Hall elements; (b) the numbers in [ ] correspond to waveform number in Fig. 5.2.

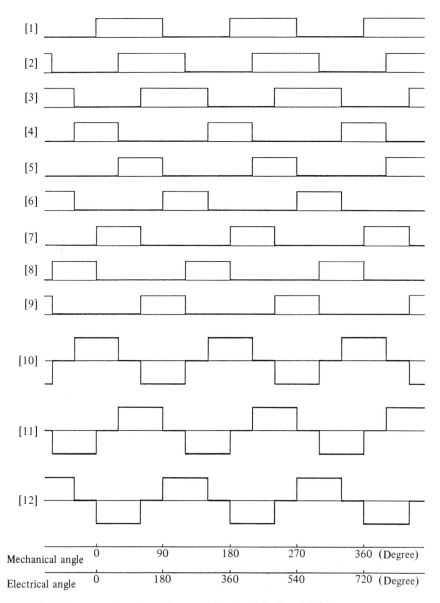

Fig. 5.2. Voltage waveform in each part of the circuit in Fig. 5.1(a).

reflected beam can be used for scanning as shown in Fig. 5.3. How an image is produced is explained, using Fig. 5.4 and the following statements:

(1) The drum has a photoconductive layer (e.g. CdS) on its surface, with photosensitivity of the layer being tuned to the wavelength of the

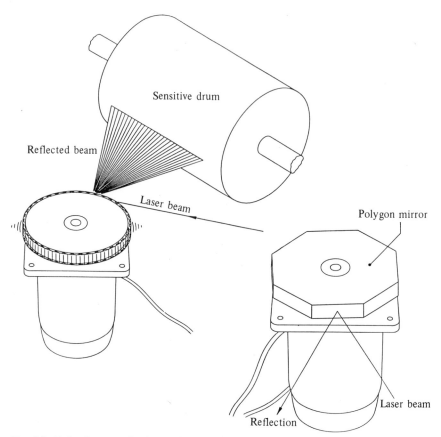

Fig. 5.3. Role of motors for laser printers; (right) a brushless DC motor driving a polygon mirror, and (above) how to scan laser beams.

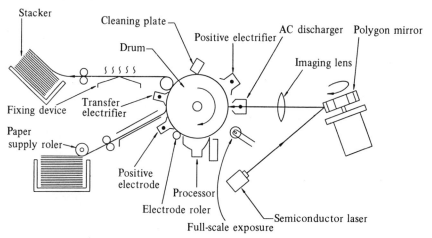

Fig. 5.4. Principles of laser printers.

Fig. 5.5. Brushless DC motor for a laser printer.

laser. The latent image of the information to be printed is formed on the drum surface by the laser and then developed by the attracted toner.

(2) The developed image is then transferred to normal paper and fixed using heat and pressure.

(3) The latent image is eliminated.

A recent brushless DC motor designed for a laser printer is shown in Fig. 5.5, and its characteristic data are given in Table 5.1.

Table 5.1. Characteristics of three-phase bipolar type brushless motors

Item	Manufacturer Model	Nippon Densan Corporation 09PF8E4036
Voltage	V	$\pm24\pm1.2$
Output	W	36
Rated torque	10^{-1} N m	0.294
Starting torque	10^{-1} N m	0.588
Starting time	s	3 (at non-inertial load)*
Rated speed	r.p.m.	6000, 9000, 12 000 selection
Rated current	A	3.5
Temperature	°C	$5\sim45$
Stability	per cent	±0.01
		Three-phase Δ connection

* A non-inertial load is a load applied by using a pulley and a weight.

5.2 Three-phase Y-connected unipolar motor

The type of brushless DC motor shown in Section 4.1 is called a Y-connected unipolar motor. In this motor, current never alternates in a

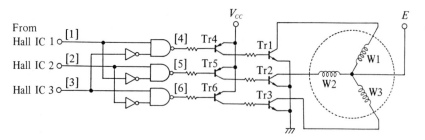

Fig. 5.6. Basic circuit diagram for a Y-connected unipolar motor.

winding. When Tr1 turns on, a current flows from the power supply E to GND through Tr1 and W3, while, if Tr1 is off no current flows. Compared with the bipolar-driven motor, the temporal utilization of each coil is one half, and the motor efficiency (the ratio of the input power to the mechanical work done by the motor) is slightly less. However, as can be seen from Figs. 5.1 and 5.6, the unipolar motor needs fewer electronic parts (such as Hall elements and transistors) and uses a simpler circuit. For these reasons unipolar-driven motors are widely used in low-cost instruments.

Figure 5.6 shows a typical, practical circuit for a Y-connected unipolar motor with three Hall ICs. A set of segment magnets are used as the rotor magnets in this class of motor to obtain as much flux as possible. Figure 5.7 shows the relation between the output signals from the Hall

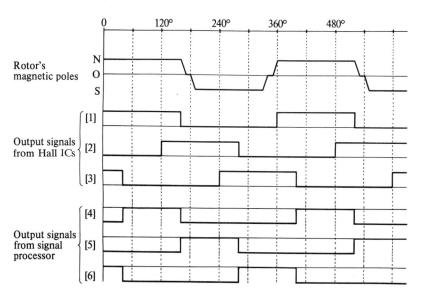

Fig. 5.7. Process of producing driving signals for Tr4 to Tr6 from the output signals of the Hall ICs.

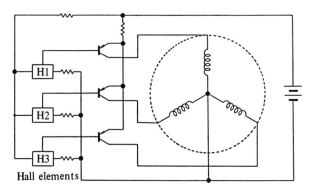

Fig. 5.8. Simplified driving circuit for a Y-connected unipolar brushless DC motor.

ICs and the switching signals for the transistors, which is implemented by logic gates.

Figure 5.8 shows a very simple driving circuit for a Y-connected unipolar motor using three Hall elements. This circuit is suited to a small motor whose output power is less than a few watts.

A typical application of brushless DC motors of this class can be found in a disk memory apparatus like a floppy disk drive system. But, due to the demand for miniaturization of memory devices, smaller and more efficient motors with higher performance have been developed. As the main secondary memory device of the computer, hard disks provide a far greater information storage capacity and shorter access times than either a magnetic tape or a floppy disk. For this reason, hard disk drives have recently been brought into use even in desktop computers. Typical disk sizes are 13.3 cm ($5\frac{1}{4}$ in.), 20 cm (8 in.) and 35.5 cm (14 in.). The number of disks varies from one to eight. Formerly, AC synchronous motors were used as the spindle motor in floppy or hard disk drives. However, brushless DC motors which are smaller and more efficient have been developed for this application and have contributed to miniaturization and increase in memory capacity in computer systems. Table 5.2 compares a typical AC synchronous motor with a brushless DC motor when they are used as the spindle motor in an 8-in. hard disk drive. As is obvious from this table, the brushless DC motor is far superior to the AC synchronous motor. Although the brushless DC motor is a little complicated structurally because of the Hall elements or ICs mounted on the stator, and its circuit costs, the merits of the brushless DC motor far outweigh the drawbacks.

The hard disk drive works as follows (see Fig. 5.9): The surface of the aluminium disk is coated with a film of magnetic material. Data is read/written by a magnetic head floating at a distance of about 0.5 μm from the disk surface. With this type, the magnetic head rises above the disk surface due to the airflow caused by the rotating disk, and this maintains a constant gap. Therefore, when the disk is stopped or slowed

Table 5.2. Comparison of an AC synchronons motor and a brushless DC motor for an 8-in. hard disk drive

	AC synchronous motor	Brushless DC motor
Power supply: direct current, low voltage (for extension and interchangeability)	Inverter required	Direct current, low voltage (12–24 V)
Speed adjustment	Since speed depends on the frequency, regional adaptability is low	Adjustable independent of frequency
Adjustment of starting time	Adjustment not possible	Adjustment possible
Temperature rise	High	Low
Efficiency	Low (approx 30 per cent)	High (40–50 per cent)
Output to volume ratio	Small (bad)	Large (good)
Speed control	Fixed	Feedback control
Structure/cost	Simple, low cost	Slightly complicated, control circuit is not so expensive by the use of ICs

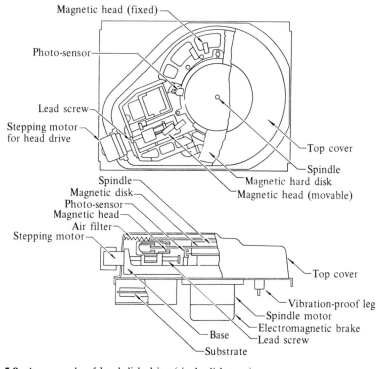

Fig. 5.9. An example of hard disk drive (single disk type).

Table 5.3. Characteristics of a three-phase unipolar motor designed for the spindle drive in a hard disk drive

Item	Manufacturer	Nippon Densan Corporation	
	Model	09FH9C4018	09FH9C4022
Voltage	V	24±2.4	24±2.4
Output	W	18	22
Rated torque	10^{-1} N m	0.490	0.588
Starting torque	10^{-1} N m	1.47	1.96
Starting time	s	1.35	1.55
Rated speed	r.p.m.	3600	3600
Rated current	A	2.0	2.4
Temperature	°C	0~50	
Stability	per cent	±1.0	
Inertia	10^{-6} kg m^2	1380	1670
Braking method		Electromagnetic method	
Number of disks		2	4

down, the head may touch the disk and cause damage to the magnetic film. To prevent this, this spindle motor must satisfy strict conditions when starting and stopping.

Table 5.3 lists the basic characteristic data of brushless DC motors used in 8-in. hard disk drives (Fig. 5.10). The starting time, which varies with the number of magnetic heads used, is about 13 seconds when four disks are loaded. The stopping time can generally be close to or less than the starting time by means of an electromagnetic brake.

Although it does not appear in Table 5.3, there is a severe restriction on the leakage flux that can affect the data stored on the disk. Hence precautions have to be taken when designing a magnetic circuit in the

Fig. 5.10. A brushless DC motor used for 8-in. hard disk drives.

motor. There are other restrictions on vibration, axis deviation, and temperature rise. Moreover the disks must be driven in dust-free conditions.

5.3   Four-phase motors

The four-phase motor, which is known as the two-phase push–pull type using two Hall elements, has been widely used since brushless motors were first used. A basic circuit is illustrated in Fig. 5.11. Hall elements H1 and H2 are placed at 90° electrical angles to each other.

If, as shown in the figure, H2 and the north pole of the rotor oppose each other, and the Hall electromotive force V_{H2} generated at output terminal 1′ turns on transistor Tr2—thereby supplying a current I_{W2} to the stator winding W2 and creating a south pole—the rotor will rotate counter-clockwise through 90°. As a result, H2 experiences no flux and loses its output voltage while H1 generates an output voltage since it opposes the south pole of the rotor. If an electromotive force V_{H3} is generated at the H1 output terminal 2, it turns on transistor Tr3, thereby supplying a current I_{W3} to the winding W3 and creating a south pole causing the rotor to rotate counter-clockwise through another 90°. Thus, the rotor will have travelled 180° from its initial position (as in Fig. 5.11) and reverses its polarity. H2 will then oppose the south pole and generate electromotive force V_{H4} causing the Tr4 transistor to be turned on, creating a south pole at winding W4 which then rotates the rotor through another 90°. In this manner the rotor revolves continuously. Figure 5.12 illustrates the relationships between Hall electromotive force V_{H2}, V_{H4} and winding currents I_{W2}, I_{W4}.

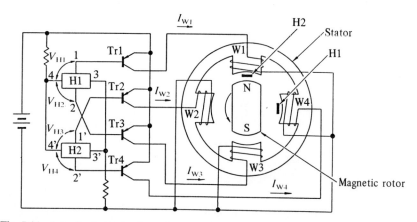

Fig. 5.11. A basic circuit for four-phase brushless DC motors.

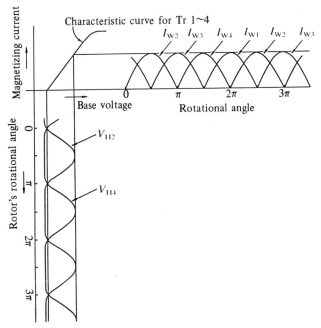

Fig. 5.12. Relation between the Hall electromotive forces (V_{H2} and V_{H4}) and currents (I_{W2} and I_{W4}); I_{W1} and I_{W3} are also shown.

5.3.1 *Electronic governor type*

Figure 5.13 exemplifies the most basic and practical type of motor using an electronic governor circuit. In this circuit, the back e.m.f. generated in the motor winding is used as a speed feedback signal. The following is a discussion of this control method.

Since in a brushless DC motor a permanent-magnet rotor revolves, a back e.m.f. proportional to the rotational speed is always generated in the windings. When the back e.m.f. from each winding is collected through diodes D1 to D4, a DC voltage having a ripple component with a frequency four times as high as the rotor's speed can be obtained. When this voltage is passed through the filter composed of the variable resistor R_2 and capacitor C, the ripple component is eliminated and converted into a DC voltage proportional to the rotational speed. This is equivalent to having a DC tachogenerator. Figure 5.14 indicates the relationship between the back e.m.f. and the rotational speed. To utilize this back e.m.f. as a speed feedback signal, a back e.m.f. comparator circuit, which is indicated by A in Fig. 5.13, is used. Moreover, the block B is added so as to decrease the total output current when the speed increases and to slow it down. The variable resistor R_1 functions as a balancer to adjust the output from the two Hall elements; R_2 controls the amount of

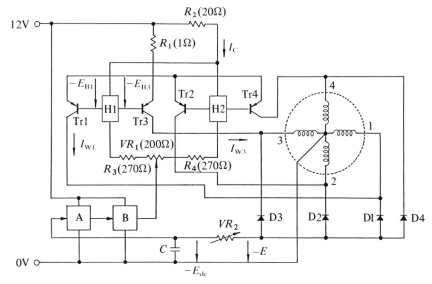

Fig. 5.13. An electronic governor type circuit designed for a four-phase motor using two Hall elements; the part marked A is the block which compares the back e.m.f. with a reference voltage, and the part marked B is the block controlling the current which flows through the Hall elements.

feedback from back e.m.f. ($-E_{dc}$) and consequently makes the rotational speed variable.

In order for a four-phase motor to have the best rotation control characteristics, the current flowing through the windings must be a correct sine wave. Further, to reduce the magnetic distortion, a cylindrical magnet is used for the rotor and a ring-shaped yoke is used for the stator core.

In this motor the outputs from the two Hall elements are sine waves which are shifted 90° from one another and each winding is supplied with a sine wave magnetizing current. If the magnetizing current which is controlled by the Hall element H1 is sin θ, then that controlled by H2 will

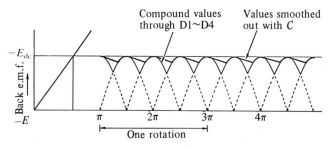

Fig. 5.14. Relations between rotational speed and back e.m.f.

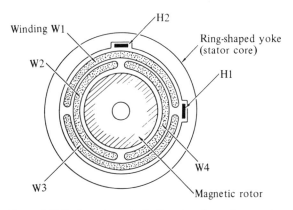

Fig. 5.15. Positions of Hall elements and windings in a four-phase motor.

be $\cos \theta$ since there is a 90° phase difference. When these two currents have the same amplitude, then the vectorially added value is constant. Hence, to give the best rotational motion, the magnitude and the angle between the magnetic field and the current vector are constants. Figure 5.15 shows a cross-sectional diagram of this type of motor.

In the motors shown in Figs. 5.11 and 5.15 there are two magnetic poles on the rotors. These motors are called two-pole motors. However, four-pole motors are also widely used. If you open a four-pole motor, you can see Hall ICs which are placed at 135° (as shown in Fig. 5.16). Since a mechanical angle of 135° is equivalent to an electrical angle of 270° in the four-pole motor, this is the same as 90° measured in the opposite direction.

Recently, brushless DC motors with their accompanying circuits have become widely used. Figure 5.17 shows an example of the appearance of a four-phase motor of this type.

Table 5.4 lists some catalogue data of four-phase (two-phase push–pull) motors.

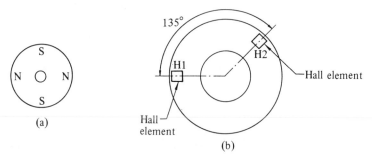

Fig. 5.16. Positions of Hall elements in a four-pole motor: (a) rotor, (b) Hall elements.

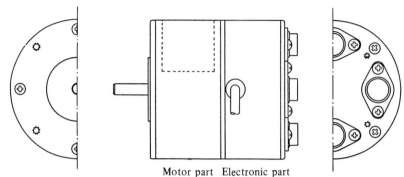

Motor part Electronic part

Fig. 5.17. Brushless DC motor with its accompanying circuits.

Table 5.4. Catalogue data for four-pole (two-phase push–pull) motors

Items	Manufacturer	Nippon Densan Corporation					
	Model	02FN8C 2002	02FYMF 2001	03FN8C 4004	03PFNF 4004	06FN8C 4015	06PNOF 4018
Voltage	V	24^{+2}_{-4}	24 ± 1	24 ± 2	12 ± 1.2	24 ± 2	24 ± 1
Output	W	1.5	0.2	4	4	15	18
Rated torque	$10^{-1}\,\mathrm{N\,m}$	0.0392	0.0196	0.147	0.0980	0.490	0.490
Starting torque	$10^{-1}\,\mathrm{N\,m}$	0.0980	0.0784	0.294	0.294	1.47	1.96
Starting time	s	0.2	0.1	1	1.3	3	3
Rated speed	r.p.m.	3600	600	3000	400–4000	3000	3600
Rated current	A	0.25	0.15	0.7	1.5	2	2
Temperature	°C	−5–+55	+20–+55	−5–+55	−10–+60	−5–+55	0–40
Stability	per cent	±1.0	±1.0	±1.0	±0.1	±1.0	±0.1

5.3.2 *Application to ultrasonic bathometers and shoal detectors*

There are many applications for the four-phase brushless DC motor and it is not possible to discuss all of them in such a limited space. As seen in Fig. 5.12, two-phase motors are equivalent to four-phase motors, and in principle, the two-phase motors have a more constant torque over a wide range of speed as compared with the motors discussed thus far. Because of this advantage, these four-phase motors are often used in instruments such as ultrasonic bathometers and shoal detectors, which require a wide range of speeds and high stability.

Bathometers and shoal detectors can measure the distance to an object (e.g. a shoal of fish) utilizing the constancy of speed of ultrasonic pulses travelling through water. Figure 5.18 shows the principles of a shoal detector. The drum in the recorder is rotated by a brushless DC motor at a constant speed. When the pen gets to the starting point of the chart paper, an ultrasonic generator is activated and emits a pulse of ultrasonic waves. The pen records a reflected wave. Therefore, the position of the fish *D* is calculated by using the distance between the starting point and

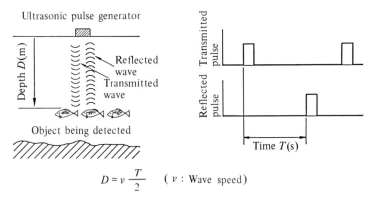

$$D = v \frac{T}{2} \quad (v : \text{Wave speed})$$

Fig. 5.18. Principles of measuring distance by ultrasonic wave.

the recorded point on the recorder. The brushless DC motor is responsible for driving the drum at a constant speed and simultaneously controlling the switching of the ultrasonic generator. The ultrasonic pulses are generated at equal intervals and the positions of the shoals are continuously recorded by monitoring these reflected waves.

5.4 Two-phase brushless motors

Since we have already seen the principles behind the two-phase motors in Section 4.5, we shall now examine the structure of some two-phase motors in practical use. There are various designs that avoid deadpoints through the use of harmonic fields.

5.4.1 *Motors with uneven airgaps*

Figure 5.19 shows a cross-sectional view of an outer-rotor two-phase motor. It has four poles and is characterized by a gap increasing in a

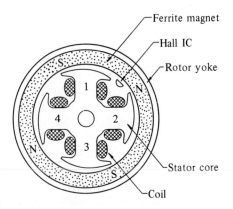

Fig. 5.19. Cross-sectional view of a motor having uneven gaps.

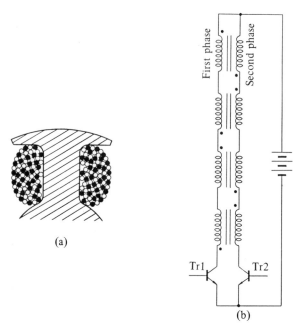

(a)

(b)

Fig. 5.20. Bifilar windings and their connections.

clockwise direction between the salient poles of the stator and the rotor magnet. This type of winding uses two coils, i.e. the first and the second phase windings which are wound together as shown in Fig. 5.20(a). The magnetic polarities of these windings are opposite to each other at each pole as shown in Fig. 5.20(b). This type of winding is called a bifilar winding.

Figure 5.19 shows the state in which the motor receives no current; the magnetic poles on the rotor are stably situated at the positions where the gaps are the narrowest.[11] This rotor position is referred to as a detent position. A clockwise torque will be created if the motor is turned slightly counter-clockwise; conversely, a slight clockwise turn will produce a counter-clockwise torque. Thus, a displacement of the rotor in either direction will produce a force in the opposite direction to restore it to the detent position. Since there are four magnetic poles on the rotor and the same number of salient poles on the stator, the relationship between the torque and the rotor angular position will be as plotted by curve (a) in Fig. 5.21. This curve has four detent positions at which the torque curve intersects the horizontal axis with a positive slope. The zero-torque positions between two adjacent detent positions are unstable positions where the rotor turns in either direction when a small external force is applied.

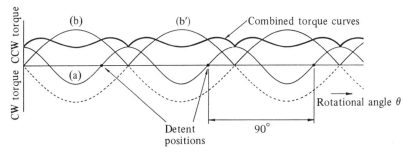

Fig. 5.21. Torque waveforms in a two-phase motor.

When the windings receive a current, the relationship between the torque and the rotational angle is as drawn by curves (b) and (b'). The zero-points of these two curves are situated somewhat to the right (CW) of the detent positions. Commutation is carried out by a group of transistors being steered by the Hall element, thus leaving the positive half of the original sine wave illustrated as (b) and (b') (see Fig. 5.21), which is the torque created by the interaction between windings and the permanent magnet. The summation of these upper halves of the curves together with the curve (a) produce the thick curve in the figure. It eliminates the deadpoints and makes the operation stable.

This method employs a simple structure, but the efficiency is lowered due to the wide gaps.

More sophisticated gap structures are discussed in reference [1].

5.4.2 *Two-phase motor having auxiliary salient poles*

Figure 5.22 shows a cross-section of the two-phase motor which belongs to this category.[2] This type of motor is characterized as follows:

(1) The rotor is magnetized N–S–O–N–S–O, where O indicates no magnetization.

(2) The stator has auxiliary salient poles with no windings.

Fig. 5.22. Two-phase motor having auxiliary salient poles; rotor is now at a stable position. (After reference [2].)

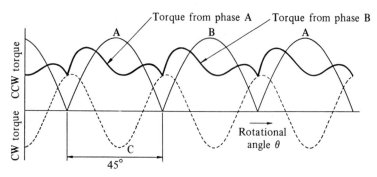

Fig. 5.23. Torque due to salient poles, torque due to excitation, and their combination.

The position of the rotor in Fig. 5.22 is close to a position at which the motor comes to rest when no current is supplied. There are eight such positions at about 45° intervals. The dotted curve C in Fig. 5.23 represents the relationship between the non-excited torque and the rotor position. Now let us look at what will happen to the torque when current is applied to the windings. If the circuit has been designed such that phase A creates the south poles on its two teeth when the rotor position is as in Fig. 5.22, the torque then becomes zero after about a 50° rotation. However, if the current is commutated to the phase B, there will again be a CCW torque, and it will continue in the same direction through 90°. The current is then switched back to phase A. The thin curve indicated by A and B in Fig. 5.23 represents this torque. Combining this torque with the reluctance torque given by the dotted sine-wave, we obtain the overall thick curve which represents a CCW torque without deadpoints.

There is, however, a problem in realizing a brushless DC motor on this simple principle. Some parts of the magnet are not magnetized; these parts will not affect the Hall element. A solution to this problem is to

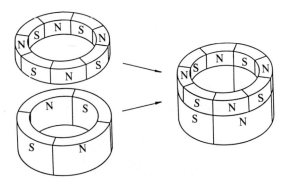

Fig. 5.24. Structure of the rotor magnets which is a combination of a four-pole and an eight-pole magnet.

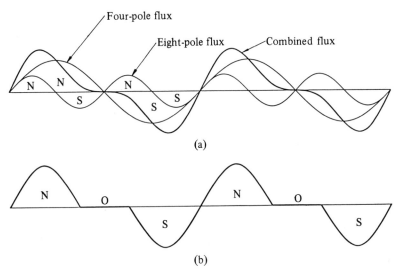

Fig. 5.25. Combination of (a) a four-pole and an eight-pole magnet producing a similar magnetic flux to (b) an NOSNOS magnetization.

combine a four-pole magnet and an eight-pole supplementary magnet so that the Hall element can sense the magnetic field produced by the four-pole magnet (see Fig. 5.24).

The result of combining the four-pole and eight-pole magnets on the N–O–S–N–O–S magnetization is illustrated in Fig. 5.25. Note the similarity between the thick curve in (a) and the curve in (b). Figure 5.26 shows a fan which employs an outer-rotor of this sort.

5.4.3 Two-phase motor having a coil pitch less than 180° electrical angle[3]

This method uses a coil pitch larger than 180° by creating a non-magnetized part on the permanent magnet.[3] Figure 5.27 shows the magnetization patterns and the positions of the coils of phase A and B in an outer-rotor motor.

When a current flows in coil segment A1 in the direction shown in Fig. 5.27, which is coming from the surface, the torque acting on the coil segment will be a function of the position as shown in Fig. 5.28(a). The figure also shows the torque acting on the coil segment A2. The sum of both of these is the torque acting on the phase A coil as shown in (b). Although positive and negative torque will alternate, the positive torque covers an area of rotational angle greater than 180°. Therefore, if the current is commutated between phases A and B such that the torque appears alternatively, the combination of phases A and B will produce a torque curve without deadpoints as illustrated in (c).

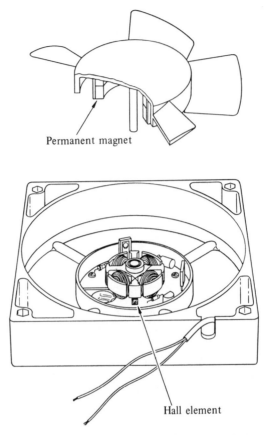

Permanent magnet

Hall element

Fig. 5.26. Two-phase brushless motor with auxiliary salient poles built into a slimline fan.

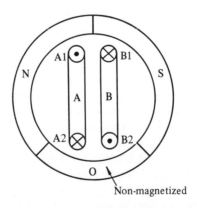

Non-magnetized

Fig. 5.27. A two-phase brushless DC motor having a coil pitch less than $180°$ electrical angle.

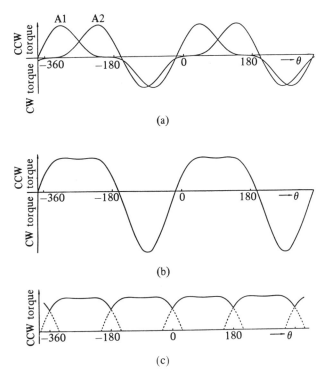

Fig. 5.28. Torque waveforms and combined curves in the two-phase motor having a coil pitch less than 180°: (a) torque acting on A1 and A2 vs. θ, (b) torque vs. θ when current flows in phase A, (c) torque when current is commutated between phases A and B at 180° intervals.

5.5 Brushless phonomotors

One application of a brushless DC motor is in a record player; its turntable is directly driven by a brushless DC motor. There are several types of motors that can be used, but only the most typical type is discussed here.[4]

Recent phonomotors are flat as illustrated in Figs. 5.29 and 5.30, and have the following features:

(1) The coils are arranged flat. They are similar to coreless DC motors in that they do not have slots to install coils as in conventional motors. Figure 5.31 illustrates a typical shape of these coils.

(2) In addition to the main magnetic poles necessary for the motor functions, the rotor magnet has fine magnetic poles which are used as a speed detector (see Fig. 5.32).

(3) The speed detector coil is found on the printed board as shown in Fig. 5.33.

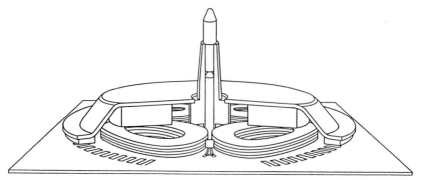

Fig. 5.29. Cutaway view of a brushless DC motor used to drive a turntable in a record player.

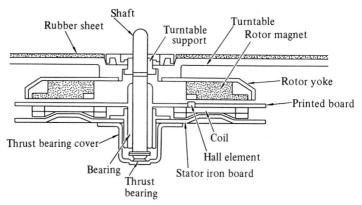

Fig. 5.30. Cross-sectional view of a phonomotor. (After Ref. [4].)

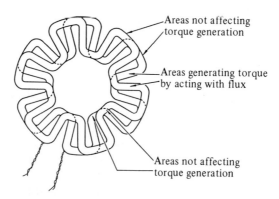

Fig. 5.31. Twelve-pole, two-phase wave windings.

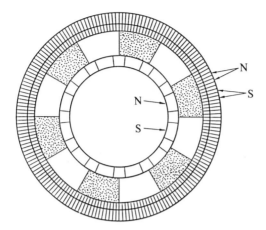

Fig. 5.32. Pole arrangement on the rotor magnet. (After Ref. [4].)

As seen in Fig. 5.31, the two windings are arranged at 90 electrical degrees from one another. If bipolar currents are supplied, and their waveforms are shaped like those in Fig. 5.34, the torque generated and the force along the shaft will show little fluctuation.

The main consideration in designing a phonomotor is to eliminate motor fluctuations and unevenness of torque and to maximize the starting torque. A large starting torque is desirable so that the motor can reach a specific rotational speed as soon as possible. After reaching the desired speed ($33\frac{1}{3}$ or 45 r.p.m.), a servo-circuit is activated so that the constant speed can be maintained with a small amount of power. Since the input power at a constant speed is as small as 2 W, the heat produced inside the motor would still be minimal even if all of the input power was converted to heat. Therefore, ordinary energy efficiency is not the first design priority; rather, a more important figure is the amount of starting torque that can be obtained per input power. This is called the 'torque efficiency'.

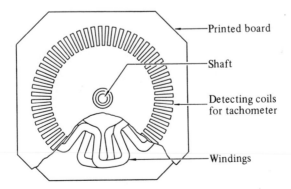

Fig. 5.33. Sensor coil on a printed board used for speed detection.

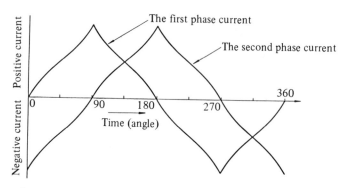

Fig. 5.34. Current waveforms on each phase.

Since the parts of the coils that contribute to torque generation are the parts which are directed outwards from the centre, as indicated in Fig. 5.31, according to the above principle, the torque efficiency should increase as the number of poles increase. However, too many magnetic poles are undesirable since the magnetic flux produced from the same area of a magnet decreases as the number of poles increase. According to the previous reference, maximum torque efficiency is obtained when the magnet has 12 poles.

References for Chapter 5

[1] Müller, R. (1975). Collector-less DC motor. US Patent 3 873 897.
[2] Wessels, J. H. (1967). Self-starting direct-current motors having no commutator. US Patent 3 299 355.
[3] Uzuka, M. (1980). DC Motors. US Patent 5 217 508.
[4] Igarashi, Y. (1980). Flat-type slotless phono motor for direct drive player [in Japanese]. *National Technical Report* **26,** (Oct.) 774–82.

6. Calculation of servomotor characteristics

The nature of a servomotor can be quantitatively expressed in terms of static and dynamic characteristics. The static characteristics indicate values of torque, efficiency, current, etc. in the stationary state. Dynamic characteristics are related to the changes of these values during operation. This chapter discusses methods for solving static characteristic problems by using equivalent circuits for DC servomotors. Methods of measuring static characteristics will be studied as well.

Dynamic characteristics will be discussed in detail in the next chapter.

6.1   Equivalent circuit and static characteristics of DC motors

As previously stated, a DC motor can be represented by an equivalent circuit shown in Fig. 6.1(a) (see also Section 1.6). Here we will use the equivalent circuit (b) to consider the losses other than those which occur in the armature resistance R_a. The theoretical grounds for this equivalent circuit are discussed below.

The Joule heat $I_a^2 R_a$ in the armature resistance R_a is called the copper (conduction) loss. Other losses producing useless heat are:

(1) *Windage loss*: work required for the rotor to move through the air. This is converted to heat inside the rotor and in the air.

(2) *Mechanical loss*: loss caused by brush and bearing friction.

(3) *Iron loss*: eddy-current and hysteresis losses caused by the rotation of the iron core of the rotor in the field flux.

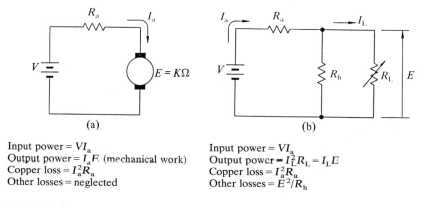

Input power = VI_a
Output power = $I_a E$ (mechanical work)
Copper loss = $I_a^2 R_a$
Other losses = neglected

Input power = VI_a
Output power = $I_L^2 R_L = I_L E$
Copper loss = $I_a^2 R_a$
Other losses = E^2/R_h

Fig. 6.1. Equivalent circuits for a DC motor.

Of these, (1) and (2) are considered to be in proportion to the square of the speed. As seen in Section 1.4, the speed is proportional to the back e.m.f., E. If a resistance R_h is placed as shown in Fig. 6.1(b), the loss in R_h is E^2/R_h. Hence the two losses can be represented by the loss in R_h by choosing an appropriate R_h. In the iron loss (3), the hysteresis loss is proportional to the speed, while the eddy-current loss is almost proportional to the square of the speed and can be included in R_h as in windage or mechanical loss. Hence, an approximate value of R_h is determined by including both components of the iron loss. In moving-coil (coreless) motors, however, the iron loss is very small.

The mechanical output is expressed by the power consumed in R_L in the equivalent circuit. A variable resistance R_L is used because the power consumed varies with the frictional torque of load T_L and with the rotational speed Ω. R_L is infinite, when no load is applied, and zero when the shaft is locked (that is, R_L ranges from 0 to ∞).

6.2 Calculation of static characteristics

Static characteristics of the DC motor can be calculated by using the
equivalent circuit in Fig. 6.1(b) and some parameters obtained by simple
measurements. The method which is presented here is based on a report
by Mas[1] but an improved one. We will now define the constant M as
follows:

$$M = \sqrt{\{(R_a + R_h)/R_a\}}. \tag{6.1}$$

To obtain the value of M from a simple measurement, we use

$$M = \sqrt{\{(\Omega_0/I_0)(\Delta I/\Delta\Omega) + 1\}} \tag{6.2}$$

where
  Ω_0 is the no-load speed (rad/s),
  $\Delta\Omega$ is the change in rotational speed when a light load is applied
(rad/s),
  I_0 is the no-load current (A), and
  ΔI is the change in current when a light load is applied (A).
Note that all of the above variables represent measured values.
  First, we want to know the maximum efficiency, or the ratio of the
mechanical output power to the input electrical power. In terms of M this
is given by

$$\text{(maximum efficiency)}\quad \eta_{max} = (M-1)/(M+1). \tag{6.3}$$

Moreover when a motor provides the maximum efficiency, the current
and speed are given by, respectively

$$\text{(current at maximum efficiency)}\quad I^* = MI_0 \tag{6.4}$$

$$\text{(speed at this point)}\quad \Omega^* = \{M/(M+1)\}\Omega_0. \tag{6.5}$$

  Proof for these five relations are as follows: First we shall prove that
eqns (6.1) and (6.2) are mathematically equivalent. When a load is applied
to a motor, the current is given by

$$I_a = (V-E)/R_a = (V-K\Omega)/R_a \tag{6.6}$$

where K is the motor constant which differs from one motor to another as
explained in Section 1.3.
  Therefore, the no-load current is

$$I_0 = (V-K\Omega_0)/R_a. \tag{6.7}$$

From eqns (6.6) and (6.7), we obtain

$$I_a - I_0 = K(\Omega_0 - \Omega)/R_a. \tag{6.8}$$

Therefore, we have

$$\frac{\Delta I}{\Delta\Omega} = \frac{I_a - I_0}{\Omega_0 - \Omega} = \frac{K}{R_a}. \tag{6.9}$$

On the other hand, since $R_L = \infty$ at no-load, we have the relation

$$\text{(no-load current)} \quad I_0 = V/(R_a + R_h). \tag{6.10}$$

The voltage E_0 that appears on R_h is

$$E_0 = R_h I_0 = V \frac{R_h}{R_a + R_h}. \tag{6.11}$$

Then the no-load speed Ω_0 is

$$\Omega_0 = E_0/K = \frac{V}{K} \cdot \frac{R_h}{R_a + R_h}. \tag{6.12}$$

Therefore,

$$\frac{\Omega_0}{I_0} = \frac{R_h}{K}. \tag{6.13}$$

Since substitution of eqns (6.9) and (6.13) for eqn (6.2) gives eqn (6.1), it has been proven that eqns (6.1) and (6.2) are theoretically equivalent.

Next, we shall prove eqn (6.3) for the maximum efficiency. By referring to the equivalent circuit, we can obtain the following three equations.

$$\text{(losses)} = \frac{(V-E)^2}{R_a} + \frac{E^2}{R_h} \tag{6.14}$$

$$\text{(output power)} = \text{(back e.m.f.} \times \text{input current} - \text{loss in } R_h)$$

$$= \frac{E(V-E)}{R_a} - \frac{E^2}{R_h} \tag{6.15}$$

$$\text{(input power)} = \text{(losses} + \text{output power)}$$

$$= \text{(terminal voltage} \times \text{input current)}$$

$$= V \frac{V-E}{R_a}. \tag{6.16}$$

Therefore, the efficiency η is

$$\eta = \frac{\text{Output power}}{\text{Input power}} = \left(\frac{E(V-E)}{R_a} - \frac{E^2}{R_h} \right) \bigg/ \left(V \frac{V-E}{R_a} \right) = \frac{E}{V} - \frac{R_a}{R_h} \cdot \frac{E^2}{V(V-E)}. \tag{6.17}$$

To find the maximum value, we differentiate η with respect to E and let it equal 0.

$$\frac{\partial \eta}{\partial E} = \frac{1}{V} \left\{ 1 - \frac{R_a}{R_h} \cdot \frac{2EV - E^2}{(V-E)^2} \right\} = 0. \tag{6.18}$$

From eqn (6.18), a quadratic equation in E is obtained as follows:

$$\left(1 + \frac{R_h}{R_a} \right) E^2 - 2V \left(1 + \frac{R_h}{R_a} \right) E + \frac{R_h}{R_a} V^2 = 0. \tag{6.19}$$

By using the relations of eqn (6.1), this is converted to an equation in E as follows:

$$M^2E^2 - 2VM^2E + (M^2 - 1)V^2 = 0. \tag{6.20}$$

Therefore, the back e.m.f. E^* that maximizes η is

$$E^* = \frac{VM^2 \pm \sqrt{(M^4V^2 - M^2(M^2-1)V^2)}}{M^2} = \frac{M^2 \pm M}{M^2}V = \frac{M \pm 1}{M}V. \tag{6.21}$$

Since V is always greater than E^* when a DC machine is operated as a motor

$$E^* = \{(M-1)/M\}V. \tag{6.22}$$

Substituting eqn (6.22) into eqn (6.17), we obtain

$$\eta_{max} = \frac{M-1}{M} - \frac{R_a}{R_h} \cdot \frac{(M-1)^2}{M}. \tag{6.23}$$

But from eqn (6.1),

$$R_h/R_a = M^2 - 1 = (M-1)(M+1). \tag{6.24}$$

Substituting this into eqn (6.23), we get

$$\eta_{max} = (M-1)/(M+1) \tag{6.25}$$

which is the same as eqn (6.3).

The current at this point is calculated by using $I_a = (V-E)/R_a$ of eqn (6.6) as follows:

$$\frac{I^*}{I_0} = \frac{(V-E^*)/R_a}{(V-E_0)/R_a} = \frac{V-E^*}{V-E_0}. \tag{6.26}$$

From eqns (6.1) and (6.11),

$$E_0 = V\frac{R_h/R_a}{(R_h/R_a)+1} = \frac{M^2-1}{M^2}V. \tag{6.27}$$

Substituting eqns (6.22) and (6.27) into eqn (6.26) we obtain

$$\frac{I^*}{I_0} = \frac{1-(M-1)/M}{1-(M^2-1)/M^2} = M. \tag{6.28}$$

Thus, eqn (6.4) has been obtained.

Also, the rotational speed at the maximum efficiency is obtained by the following calculations. Let us start from the following relation:

$$\frac{\Omega^*}{\Omega_0} = \frac{E^*}{E_0}. \tag{6.29}$$

Substituting eqns (6.27) and (6.22) into eqn (6.29), we obtain

$$\frac{\Omega^*}{\Omega_0} = \frac{V(M-1)/M}{V(M^2-1)/M^2} = \frac{M}{M+1} \tag{6.30}$$

and thus, eqn (6.5) is obtained.

Finally, the torque T^* at the maximum efficiency can be derived as follows:

$$
\begin{aligned}
T^* &= (\text{input power} \times \eta_{max}/\Omega^*) \\
&= VI^*\eta_{max}/\Omega^* \\
&= VMI_0\{(M-1)/(M+1)\}/\{\Omega_0 M/(M+1)\} \\
&= (M-1)VI_0/\Omega_0. \tag{6.31}
\end{aligned}
$$

Table 6.1 summarizes these results, including the input and output power.

Table 6.1. Expressions of the maximum efficiency and other quantities at this value

Coefficient M	$\sqrt{((\Omega_0/I_0)(\Delta I/\Delta\Omega)+1)}$
Maximum efficiency η_{max}	$(M-1)/(M+1)$

The following values are at the maximum efficiency:

Current I^*	MI_0
Rotational speed Ω^*	$\{M/(M+1)\}\Omega_0$
Input power P_{in}^*	MVI_0
Output power P_{out}^*	$\{M(M-1)/(M+1)\}VI_0$
Torque T^*	$(M-1)VI_0/\Omega_0 (\text{N m}) = 10\,200 \times (M-1)VI_0/\Omega_0(\text{g cm})$

Ω_0 = no-load speed; usually expressed in r.p.m. (revolutions per minute), but values in rad s^{-1} should be used for calculations
I_0 = no-load current

6.3 Example

Problem 1 A motor without a load is supplied with 12 V, and rotated at 4094 r.p.m. with a current of 28.2 mA. When a light load is added as in Fig. 6.2, the rotational speed changes to 4009 r.p.m. and the current is increased to 99 mA. What is the maximum efficiency of this motor? Also, calculate the input current I^*, rotational speed Ω^*, input power P_{in}^*, output power P_{out}^*, and torque T^* at this point.

Answers and explanations

$$\Delta\Omega = 4094 - 4009 = 85 \text{ r.p.m.}$$
$$\Delta I = 0.099 - 0.0282 = 0.0708 \text{ A}$$
$$M = \sqrt{\left(\frac{4094}{0.0282} \times \frac{0.0708}{85} + 1\right)} = 11.042.$$

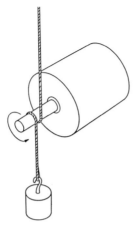

Fig. 6.2. A motor with a light load being applied.

Therefore,
$$\eta_{max} = (11.042-1)/(11.042+1) = 0.834 = 83.4 \text{ per cent}$$

where
$$I^* = 11.042 \times 0.0282 = 0.311 \text{ A},$$

$$\Omega^* = \frac{11.042}{11.042+1} \times 4094 = 3754 \text{ r.p.m.},$$

$$P_{in}^* = 11.042 \times 12 \times 0.0282 = 3.737 \text{ W},$$

$$P_{out}^* = \{11.042 \times (11.042-1)/(11.042+1)\} \times 12 \times 0.0282$$
$$= 3.116 \text{ W},$$

and
$$T^* = (11.042-1) \times 12 \times 0.0282/(4094 \times 0.1047) = 7.926 \times 10^{-3} \text{ N m}$$
$$= 10200 \times 7.926 \times 10^{-3} = 80.85 \text{ g cm}.$$

Note: The rotational speed Ω in the SI unit is obtained from the following relation:
$$1 \text{ r.p.m.} = 6.28/60 = 0.1047 \text{ rad s}^{-1}.$$

Problem 2 When the motor of problem 1 was measured with a voltage around 12 V, the results in Table 6.2 were obtained. Why do M and η_{max} vary with the different voltages?

Table 6.2. Values of M and η_{max} at various voltages

V (V)	I_0 (mA)	Ω_0 (r.p.m.)	I_a (mA)	Ω (r.p.m.)	M	η_{max} (per cent)
10	26.5	3408	96	3328	10.617	82.8
11	27.4	3753	98	3663	10.414	82.5
12	28.2	4094	99	4009	11.042	83.4
13	39.2	4440	101	4360	11.725	84.3
14	30.3	4783	103	4692	11.274	83.7

Answers and explanations   The circuit in Fig. 6.1(b) is an approximate equivalent circuit and not an exact one. The factors that render this circuit an approximation are:

(1) The voltage drop across the brushes is not considered. Precious metal brushes produce a small voltage drop, and therefore the equivalent circuit will not show much of a discrepancy. On the other hand, carbon brushes might cause a large discrepancy.

(2) The friction between the brushes and commutator consists of two portions as illustrated in Fig. 6.3; the portion independent of speed and the portion proportional to it. These two types of friction are added and represented approximately by R_h. In this equivalent circuit, R_h should be considered to vary with the terminal voltage. The ratio of the friction independent of speed to the total friction is large at higher voltages. This may be the main reason for the decrease in η_{max} with a lower voltage.

(3) As previously mentioned, hysteresis loss cannot be exactly represented by R_h. The loss decreases as the rotational speed decreases, but in proportion to the total iron loss it increases. Therefore, this may be the reason for the lower maximum efficiency with the lower terminal voltage.

As seen here, R_h is a function of the motor's terminal voltage. This means that the maximum efficiency must be indicated for the rated voltage.

Problem 3   When the no-load speed is Ω_0, what is the speed Ω_p which maximizes the output power, the maximum output P_{max}, and the efficiency η_p at that speed?

Answers and explanations   Let us employ an approximation which assumes $R_L = \infty$. The output P_{out} at this point is given by

$$P_{out} = \frac{E(V-E)}{R_a} \tag{A1}$$

Differentiating eqn (A1) with respect to E, we get

$$\frac{\partial P_{out}}{\partial E} = \frac{1}{R_a}(V - 2E). \tag{A2}$$

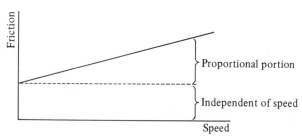

Fig. 6.3. Composition of frictional load: the part independent of speed and the other part proportional to speed.

By letting eqn (A2) equal 0, we obtain

$$E = V/2. \tag{A3}$$

This means that the output power is maximized at the speed which makes the back e.m.f. a half of the terminal voltage.

In this approximation, the no-load speed Ω_0 is given by

$$\Omega_0 = V/K. \tag{A4}$$

It is seen that eqn (A3) is satisfied when the rotational speed drops to half the no-load speed. Therefore, the rotational speed at the maximum output power Ω_p is

$$\Omega_p = \Omega_0/2. \tag{A5}$$

By substituting eqn (A3) into eqn (A1), we obtain for the maximum output P_{max}:

$$P_{max} = \frac{(V/2) \times (V - V/2)}{R_a} = \frac{V^2}{4R_a}. \tag{A6}$$

To derive the efficiency we substitute eqn (A3) into eqn (6.17) and put $R_h = \infty$, thus:

$$\eta_p = \frac{V/2}{V} - 0 = 0.5 = 50 \text{ per cent.} \tag{A7}$$

As can be seen from this example, when a DC motor is driven with a given voltage and rotating at the maximum output speed, it will have an efficiency of 50 per cent at best, and usually less. Thus, use of DC motors with such performance is undesirable (see Fig. 6.6).

Problem 4 The no-load speed of a motor is 5000 r.p.m. and the value of M obtained from an experiment is 10. What is the speed Ω^* which maximizes the efficiency?

Answer and explanation

$$\Omega^* = \{M/(M+1)\}\Omega_0 = \{10/(10+1)\} \times 5000 = 4545 \text{ r.p.m.}$$

The maximum efficiency is obtained at a speed slightly lower than the no-load speed. In general, the higher the efficiency the closer the maximum efficiency speed is to the no-load speed.

6.4 Static characteristic curves

There are two methods to express static characteristic curves:

(1) To express the efficiency η, input current I_a, input power P_{in}, output P_{out}, and torque T as functions of the rotational speed Ω.

(2) To express the rotational speed Ω, efficiency η, input current I_a, input power P_{in}, and output P_{out} as functions of the torque T.

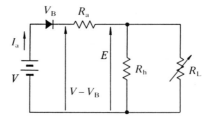

Fig. 6.4. Equivalent circuit in consideration of the voltage drop across brushes.

We shall study both methods here. Before these, however, let us first consider the voltage drop and power loss at the brushes. As explained in Section 2.6, a voltage drop of almost a constant value appears across the brushes while a current is passing through them. This property is often simulated by a semiconductor rectifier diode. Therefore, the voltage drop across the brushes can be represented by an equivalent diode as shown in Fig. 6.4. That is, the voltage drop across the brushes is approximated by the forward voltage V_B of the diode. This adds another loss $I_a V_B$.

The definition of M given by eqn (6.1) or (6.2) remains the same with this equivalent circuit.

Recently, the widespread use of personal and office computers has made it possible for many people to do computations using the computer language called BASIC. Therefore, when developing theories, we will use formulae which lend themselves to use in these computers—unnecessary simplifications will not be made.

6.4.1   Obtaining R_a, R_h, and V_B

The theoretical formulae for R_a and R_h are obtained as follows. First, when no load is applied to the motor, R_L in Fig. 6.4 is infinite. Hence we have for the voltage equation,

$$V - V_B = I_0(R_a + R_h) \tag{6.32}$$

which leads to

$$R_h = (V - V_B)/I_0 - R_a. \tag{6.33}$$

R_a and V_B are needed for the calculation and can be obtained from a lock-load test (refer to Section 6.5.4).

Since $R_L = 0$ when the motor shaft is locked,

$$V = V_B + I_a R_a. \tag{6.43}$$

If the graph in Fig. 6.5 (which shows the relationship between the applied voltage V and the armature current I_a) is obtained when a test motor is lock-loaded, then we can state:

(1) the intercept on the ordinate gives V_B, and
(2) the slope of the graph gives R_a.

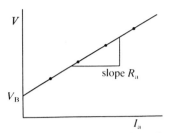

Fig. 6.5. A method to obtain V_B and R_a from the lock-load test.

When M is used to obtain the relationship between R_a, V_B, and I_0, we obtain from eqns. (6.32) and (6.1) the relation

$$R_a = \frac{V - V_B}{I_0 M^2}.$$

(6.35)

However, we shall use the R_a which is obtained from the lock-load test in later calculations. Now we are ready to develop a theory on static characteristics.

6.4.2 *Expression of quantities as functions of rotational speed*

Method 1   Substituting $V - V_B$ for V in eqn (6.14) and considering the loss at V_B gives

$$\text{losses: } P_{\text{loss}} = \frac{\{(V - V_B) - E\}^2}{R_a} + \frac{E^2}{R_h} + V_B I_a.$$

(6.36)

On the other hand,

$$I_a = (V - V_B - E)/R_a.$$

(6.37)

Substituting this expression into eqn (6.36) gives

$$P_{\text{loss}} = \frac{V - V_B - E}{R_a}(V - E) + \frac{E^2}{R_h}.$$

(6.38)

Substituting $V - V_B$ for V in eqns (6.15) and (6.16) gives

$$P_{\text{out}} = \frac{E(V - V_B - E)}{R_a} - \frac{E^2}{R_h}$$

(6.39)

$$P_{\text{in}} = V \frac{V - V_B - E}{R_a}.$$

(6.40)

Therefore, the efficiency η becomes

$$\eta = \left(\frac{E(V - V_B - E)}{R_a} - \frac{E^2}{R_h} \right) \Big/ \left(V \frac{V - V_B - E}{R_a} \right)$$

$$= \frac{E}{V} - \frac{R_a}{R_h} \cdot \frac{E^2}{V(V - V_B - E)},$$

(6.41)

and the torque T is

$$T = P_{out}/\Omega. \tag{6.42}$$

By substituting eqn (6.39) and

$$\Omega = E/K \tag{6.43}$$

into eqn (6.42), we obtain for T:

$$T = \left(\frac{V - V_B - E}{R_a} - \frac{E}{R_h}\right)K. \tag{6.44}$$

In order to express these formulae in terms of Ω, we should substitute

$$E = K\Omega \tag{6.45}$$

into each equation, where K is expressed as

$$K = (V - V_B - R_a I_0)/\Omega_0 \tag{6.46}$$

which is obtained by substituting eqn (6.45) into eqn (6.37). Since the numerator is $R_h I_0$, K is also given as

$$K = R_h I_0/\Omega_0. \tag{6.47}$$

Based on the above theories, Table 6.3 summarizes the procedure for drawing characteristic curves from simple measurements. An example program in BASIC (refer to Table 6.4) and the results of executing this program are given (refer to Table 6.5). Figure 6.6 shows the characteristic curves drawn from the data in Table 6.5.

Table 6.3. Calculation procedures to obtain characteristic curves (Method 1)

Procedure	
	Obtain the following data from no-load and lock-load test:
1	V, V_B, R_a, I_0, Ω_0
2	Obtain R_h; $R_h = (V - V_B)/I_0 - R_a$
3	Obtain K; $K = R_h I_0/\Omega_0$
4	Put $\Omega = 0$
5	Obtain E; $E = K\Omega$
6	Obtain I_a; $I_a = (V - V_B - E)/R_a$
7	Obtain P_{out}; $P_{out} = E\dfrac{V - V_B - E}{R_a} - \dfrac{E^2}{R_h}$
8	Obtain P_{in}; $P_{in} = V\dfrac{V - V_B - E}{R_a}$
9	Obtain η; $\eta = \dfrac{P_{out}}{P_{in}}$
10	Obtain T; $T = \left(\dfrac{V - V_B - E}{R_a} - \dfrac{E}{R_h}\right)K$
11	Add a certain change of $\Delta\Omega$ to the value of Ω. $\Delta\Omega$ should be of enough value to draw a graph (e.g. when Ω_0 is 4000 r.p.m., $\Delta\Omega$ should be about 100 r.p.m.).
12	Calculation is completed when $\Omega > \Omega_0$.
13	Return to procedure 5.

Note. The conventional unit for rotational speed Ω is r.p.m. but here rad s^{-1} is used as the SI unit

Table 6.4. Calculation program for static characteristic curves (Method 1)

```
100 '******  PROGRAM  1  ******
110 '
120 '  DATA
130 '
140 V = 12                        '[ V  ]
150 VB = 0                        '[ V  ]
160 RA = 3.35                     '[ Ohm ]
170 IO = .0282                    '[ A ]
180 NO = 4094                     '[ RPM ]
190 DN = 100                      '[ RPM ]
200 '
210 '  CALCULATE
220 '
230 PRINT "N [ RPM ]";SPC(4);"Ia [ A ]";SPC(4);"Pin [ W ]";SPC(4);
240 PRINT "Pout [ W ]";SPC(4);"ETA [ % ]";SPC(4);"T [ Nm ]"
250 RH = (V-VB)/IO-RA             '[ Ohm ]
260 K = RH*IO/(NO/9.549)          '[ Vs/rad ]
270 FOR N=0 TO NO STEP DN
280 OM = N/9.549                  '[ rad/s ]
290 E = K*OM                      '[ V ]
300 IA = (V-VB-E)/RA              '[ A ]
310 POUT = E*(V-VB-E)/RA-E*E/RH   '[ W ]
320 PIN = V*(V-VB-E)/RA           '[ W ]
330 ETA = POUT/PIN*100            '[ % ]
340 T = K*(V-VB-E)/RA-K*E/RH      '[ Nm ]
350 PRINT USING " ####       #.###        ##.###       ";N,IA,PIN;
360 PRINT USING "  ##.###       ##.#       #.####";POUT,ETA,T
370 IF N=NO THEN 410
380 NEXT
390 N=NO
400 GOTO 280
410 END
```

Table 6.5. Results of computation

N [RPM]	Ia [A]	Pin [W]	Pout [W]	ETA [%]	T [Nm]
0	3.582	42.985	0.000	0.0	0.0995
100	3.495	41.943	1.016	2.4	0.0970
200	3.408	40.902	1.982	4.8	0.0946
300	3.322	39.860	2.896	7.3	0.0922
400	3.235	38.818	3.760	9.7	0.0898
500	3.148	37.777	4.572	12.1	0.0873
600	3.061	36.735	5.334	14.5	0.0849
700	2.974	35.693	6.045	16.9	0.0825
800	2.888	34.652	6.705	19.3	0.0800
900	2.801	33.610	7.314	21.8	0.0776
1000	2.714	32.568	7.872	24.2	0.0752
1100	2.627	31.527	8.380	26.6	0.0727
1200	2.540	30.485	8.836	29.0	0.0703
1300	2.454	29.443	9.242	31.4	0.0679
1400	2.367	28.401	9.597	33.8	0.0655
1500	2.280	27.360	9.900	36.2	0.0630
1600	2.193	26.318	10.153	38.6	0.0606
1700	2.106	25.276	10.355	41.0	0.0582
1800	2.020	24.235	10.506	43.4	0.0557
1900	1.933	23.193	10.607	45.7	0.0533
2000	1.846	22.151	10.656	48.1	0.0509
2100	1.759	21.110	10.655	50.5	0.0484
2200	1.672	20.068	10.602	52.8	0.0460
2300	1.586	19.026	10.499	55.2	0.0436
2400	1.499	17.985	10.345	57.5	0.0412
2500	1.412	16.943	10.140	59.8	0.0387
2600	1.325	15.901	9.884	62.2	0.0363
2700	1.238	14.860	9.577	64.4	0.0339
2800	1.151	13.818	9.219	66.7	0.0314
2900	1.065	12.776	8.810	69.0	0.0290
3000	0.978	11.734	8.351	71.2	0.0266
3100	0.891	10.693	7.840	73.3	0.0242
3200	0.804	9.651	7.279	75.4	0.0217
3300	0.717	8.609	6.667	77.4	0.0193
3400	0.631	7.568	6.004	79.3	0.0169
3500	0.544	6.526	5.290	81.1	0.0144
3600	0.457	5.484	4.525	82.5	0.0120
3700	0.370	4.443	3.709	83.5	0.0096
3800	0.283	3.401	2.843	83.6	0.0071
3900	0.197	2.359	1.925	81.6	0.0047
4000	0.110	1.318	0.957	72.6	0.0023
4094	0.028	0.338	0.000	0.0	0.0000

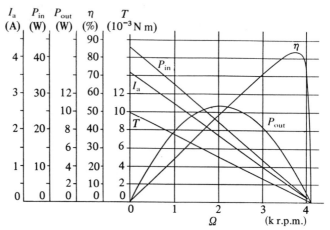

Fig. 6.6. An example of characteristic curves expressing each quantity as a function of speed.

Table 6.6. Variables and their meanings as used in programs

Program variables	Symbols and meanings used in the theory
V	V (V); Terminal voltage
VB	V_B (V); Voltage drop across brushes
RA	R_a (Ω); Armature resistance
I0	I_0 (A); No-load current
N0	Ω_0 (r.p.m.); No-load speed
DN	A change in Ω which has been added to obtain Table 5.5
RH	R_h(Ω); Loss resistance other than copper loss
K	K (V s rad^{-1}); Constant
N	Ω (r.p.m.); Rotational speed in r.p.m.
OM	Ω (rad s^{-1}); Rotational speed in the SI unit
E	E (V); Back e.m.f.
IA	I_a (A); Armature current
POU	P_{out} (W); Output power
PIN	P_{in} (W); Input power
ETA	η (per cent); Efficiency
T	T (N m); Torque

We will not discuss the program presented in Table 6.4, but Table 6.6 explains the program variables and their relationship to the symbols used in the theories.

6.4.3 Expression of quantities as functions of torque T

Method 2 The torque T equals the product of the current flowing through R_L in the equivalent circuit and the factor K:

$$T = (I_a - E/R_h)K. \tag{6.48}$$

On the other hand, since I_a is given by eqn (6.37), we get

$$T = \left(\frac{V - V_B - E}{R_a} - \frac{E}{R_h}\right)K. \tag{6.49}$$

Substituting eqn (6.43) into the above equation gives

$$T = \left(\frac{V - V_B - K\Omega}{R_a} - \frac{K\Omega}{R_h}\right)K. \tag{6.50}$$

Therefore, Ω as a function of T is:

$$\Omega = \frac{(V - V_B)R_h - (R_a R_h / K)T}{K(R_a + R_h)}. \tag{6.51}$$

The output P_{out} is easily calculated from

$$P_{out} = T\Omega. \tag{6.52}$$

Substituting eqn (6.43) into eqn (6.40) we obtain the input power P_{in} as:

$$P_{in} = V\frac{V - V_B - K\Omega}{R_a}. \tag{6.53}$$

Since eqns (6.52) and (6.53) are functions of Ω, they can be transformed into functions of T by means of eqn (6.51).

Efficiency is given as P_{out}/P_{in}, but it is not necessary to show this as a function of T.

Table 6.7 summarizes a procedure for drawing characteristic curves based on the above theories. A sample BASIC program and its results are

Table 6.7. Calculation procedures to obtain characteristic curves (Method 2)

Procedure	
	Obtain the following data from no-load and lock-load tests
1	$V, V_B, R_a, I_0, \Omega_0$
2	Obtain R_h; $R_h = (V - V_B)I_0 - R_a$
3	Obtain K; $K = R_h I_0 / \Omega_0$
4	Obtain T_{max}; $T_{max} = K\dfrac{V - V_B}{R_a}$
5	Put $T = 0$
6	Obtain Ω; $\Omega = \dfrac{(V - V_B)R_h - (R_a R_h / K)T}{K(R_a + R_h)}$
7	Obtain E; $E = K\Omega$
8	Obtain I_a; $I_a = (V - V_B - E)/R_a$
9	Obtain P_{out}; $P_{out} = T\Omega$
10	Obtain P_{in}; $P_{in} = V\dfrac{V - V_B - E}{R_a}$
11	Obtain η; $\eta = \dfrac{P_{out}}{P_{in}}$
12	Add a certain change of ΔT to the value of T. ΔT should be of proper value to draw a graph (e.g. when T_{max} is 0.1 N m, $\Delta T = 0.0025$ N m).
13	Calculation is completed when $T > T_{max}$.
14	Return to Procedure 6.

Table 6.8. Calculation program for static characteristic curves (Method 2)

```
100 '******  PROGRAM  2  ******
110 '
120 '  DATA
130 '
140 V = 12                              '[ V ]
150 VB = 0                              '[ V ]
160 RA = 3.35                           '[ Ohm ]
170 IO = .0282                          '[ A ]
180 NO = 4094                           '[ RPM ]
190 DT = .0025                          '[ Nm ]
200 '
210 '  CALCULATE
220 '
230 PRINT "T [ Nm ]";SPC(6);"Ia [ A ]";SPC(6);"Pin [ W ]";SPC(6);
240 PRINT "Pout [ W ]";SPC(6);"ETA [ % ]";SPC(6);"N [ RPM ]"
250 RH = (V-VB)/IO-RA                   '[ Ohm ]
260 K = RH*IO/(NO/9.549)                '[ Vs/rad ]
270 TM = (V-VB)*K/RA
280 FOR T=0 TO TM STEP DT
290 OM = ((V-VB)*RH-(RA*RH/K)*T)/K/(RA+RH)   '[ rad/s ]
300 E = K*OM                            '[ V ]
310 IA = (V-VB-E)/RA                    '[ A ]
320 POUT = T*OM                         '[ W ]
330 PIN = V*(V-VB-E)/RA                 '[ W ]
340 ETA = POUT/PIN*100                  '[ % ]
350 N = OM*9.549                        '[ RPM ]
360 PRINT USING " #.#####        #.###         ##.###       ";T,IA,PIN;
370 PRINT USING "   ##.###        ##.#         ####";POUT,ETA,N
380 IF T=TM THEN 420
390 NEXT
400 T=TM
410 GOTO 290
420 END
```

Table 6.9. Results of computation

T [Nm]	Ia [A]	Pin [W]	Pout [W]	ETA [%]	N [RPM]
0.00000	0.028	0.338	0.000	0.0	4094
0.00250	0.118	1.410	1.045	74.1	3991
0.00500	0.207	2.482	2.036	82.0	3888
0.00750	0.296	3.554	2.973	83.7	3785
0.01000	0.385	4.626	3.856	83.4	3682
0.01250	0.475	5.698	4.686	82.2	3580
0.01500	0.564	6.769	5.461	80.7	3477
0.01750	0.653	7.841	6.183	78.9	3374
0.02000	0.743	8.913	6.851	76.9	3271
0.02250	0.832	9.985	7.465	74.8	3168
0.02500	0.921	11.057	8.025	72.6	3065
0.02750	1.011	12.129	8.531	70.3	2962
0.03000	1.100	13.200	8.983	68.0	2859
0.03250	1.189	14.272	9.381	65.7	2756
0.03500	1.279	15.344	9.726	63.4	2653
0.03750	1.368	16.416	10.016	61.0	2551
0.04000	1.457	17.488	10.253	58.6	2448
0.04250	1.547	18.560	10.436	56.2	2345
0.04500	1.636	19.632	10.565	53.8	2242
0.04750	1.725	20.703	10.640	51.4	2139
0.05000	1.815	21.775	10.661	49.0	2036
0.05250	1.904	22.847	10.629	46.5	1933
0.05500	1.993	23.919	10.542	44.1	1830
0.05750	2.083	24.991	10.402	41.6	1727
0.06000	2.172	26.063	10.208	39.2	1625
0.06250	2.261	27.134	9.959	36.7	1522
0.06500	2.351	28.206	9.657	34.2	1419
0.06750	2.440	29.278	9.301	31.8	1316
0.07000	2.529	30.350	8.892	29.3	1213
0.07250	2.618	31.422	8.428	26.8	1110
0.07500	2.708	32.494	7.910	24.3	1007
0.07750	2.797	33.565	7.339	21.9	904
0.08000	2.886	34.637	6.714	19.4	801
0.08250	2.976	35.709	6.035	16.9	698
0.08500	3.065	36.781	5.302	14.4	596
0.08750	3.154	37.853	4.515	11.9	493
0.09000	3.244	38.925	3.674	9.4	390
0.09250	3.333	39.996	2.779	6.9	287
0.09500	3.422	41.068	1.831	4.5	184
0.09750	3.512	42.140	0.828	2.0	81
0.09947	3.582	42.985	0.000	0.0	0

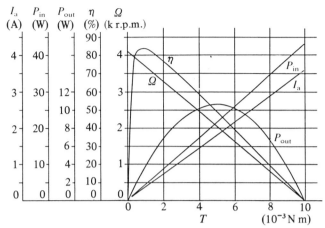

Fig. 6.7. An example of characteristic curves expressing each quantity as a function of torque T (Method 2).

given in Tables 6.8 and 6.9, respectively. The characteristic curves are shown in Fig. 6.7.

The program variables in Table 6.8 are the same as those in Table 6.6 with the following exceptions:

(1) DT is a change in torque T measured in (N m), which has been added in order to obtain Table 6.9.

(2) TM is the maximum torque measured in (N m).

Example problem As seen from the above programs, the data supplied (R_a, I_0, etc.) are the same as in Problem 1 in Section 6.3. The maximum efficiency derived in the answers for this this problem was 83.4 per cent, while it was about 83.7 per cent in Tables 6.5 and 6.9. Why is there such a discrepancy?

Answers and explanations The methods used to obtain M were different. First the definition of M used in Problem 1 of Section 6.3 was

$$M = \sqrt{((\Omega_0/I_0)(\Delta I/\Delta \Omega) + 1)} \qquad ((6.2))$$

and this was calculated from the data of speed and current measured at a no-load test and a light-loaded test.

On the other hand, the definition of M in Table 6.5 is

$$M = \sqrt{((R_a + R_h)/R_a)} \qquad ((6.1))$$

and this value is calculated from the data of a no-load and a lock-loaded test.

As explained in Section 6.2, eqns (6.2) and (6.1) are theoretically equivalent. However, one can assume that there were losses (by causes not yet discussed) depending on the kind of measurement that affects the value of $\eta_{\max}$. Thus, the previous theory for M was only an approximate one.

We shall now calculate $\eta_{\max}$ using method 2, bearing in mind that method 1 yields a value of 83.4 per cent.

Substituting eqn (6.33) into eqn (6.1) gives

$$M = \sqrt{\left(\frac{V - V_B}{R_a I_0}\right)}.$$

By substituting into this equation $V = 12\,\text{V}$, $V_B = 0\,\text{V}$, $I_0 = 0.0282\,\text{A}$ and $R_a = 3.35\,\Omega$ we get for M

$$M = \sqrt{\left(\frac{12}{3.35 \times 0.0282}\right)} = 11.271.$$

Therefore, the value obtained for $\eta_{\max}$ from a no-load test and lock-loaded test is determined from eqn (6.3) as:

$$\eta_{\max} = \frac{11.271 - 1}{11.271 + 1} = 0.837 = 83.7 \text{ per cent.}$$

6.5   Notes for installation concerning static characteristics

Up to now we have studied theories of static characteristics but now we need further knowledge in order to deal with practical problems. We will now look at some of these problems.

6.5.1   *Ratings*

The ratings stated on a motor indicate the assured limitations when driving that motor. These ratings often specify the continuous torque, rotational speed, or output power as shown in Table 3.3 in Chapter 3. However, there are three kinds of rating as explained below:

(1) *Continuous rating*: When operating within these ratings the motor will not show a temperature rise greater than the amount stated in the specifications. The motor will stay within the other limitations when used under the specified conditions.

(2) *Short-time rating*: When operating within these ratings the motor will not show a temperature rise exceeding the limit. The motor will not exceed other limitations when used for short periods after starting cold under the specified conditions.

(3) *Repetitive rating*: When operating within these ratings the motor will not show a temperature rise exceeding the limit. It will also not exceed other limitations when used repetitively in cycles alternating between a constant load and stoppages under the specified conditions.

Most catalogues list continuous ratings.

6.5.2 Problems concerning temperature

During operation, heat caused by internal losses raises the temperature of the motor. If a constant load is applied after the motor is started, the temperature will rise almost exponentially and then level off at a certain value as shown in Fig. 6.8. The difference between the temperature of the instrument and the ambient temperature is the temperature rise. When the temperature rise is too high it speeds up the deterioration of insulation, causes burns, and can cause damage to the bearings and the commutators. In general, the limits of the temperature rise are determined by the type of insulation in the motor (see Table 6.10).

The type of insulation in Table 6.10 is classified according to allowable

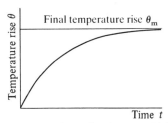

Fig. 6.8. How temperature rises with time; the time at which the temperature reaches 63 per cent of the final value is called the thermal time constant.

Table 6.10. Limits of temperature rise (°C)

Class	Measuring method	Stator winding	Armature winding
A	(1)	50	50
	(2)	60	—
	(3)	60	—
E	(1)	65	65
	(2)	75	—
	(3)	75	—
B	(1)	70	70
	(2)	80	—
	(3)	80	—
F	(1)	85	85
	(2)	100	—
	(3)	100	—
H	(1)	105	105
	(2)	125	—
	(3)	125	—

Measuring methods:
  (1) Thermometer method
  (2) Resistance method
  (3) Embedded thermometer method
  Note. For totally enclosed type, temperature limit is set 5 °C higher than these values

Table 6.11. Types of insulation and allowable maximum temperatures

Classes of insulation	Allowable maximum temperature (°C)
A	105
E	120
B	130
F	155
H	180
C	Higher than 180

safe temperatures as in Table 6.11. Examples of insulator materials are:

Class A insulation: cotton, silk, paper, polyvinyl formal.
Class E insulation: enamel or polyester film.
Class B insulation: mica, glass fibre with a conventional adhesive.
Class F insulation: glass fibre, etc. with heat-resistant adhesive.
Class H insulation: glass fibre, etc. with a silicon resin or higher grade adhesive; polyimide enamel, polyimide film, and polyamide paper are also used.
Class C insulation: ceramics.

Insulators deteriorate for one reason or another. Three main causes for this deterioration are:

(1) *Deterioration due to heat*: This is the most prominent type of deterioration. It is known that the relation between the temperature and the life of the insulator is approximated by the following equation.

$$L = ae^{-m\theta} \tag{6.54}$$

where, L is the life,
θ is the temperature of the insulator (°C),
a is a constant which depends on the material,
m is a constant which depends on the material, and
$e = 2.718$.

(2) *Deterioration due to humidity*: When water is absorbed by the insulator surface, the surface resistance coefficient decreases and the current leakage increases. Also, when water is absorbed by a dielectric substance, the bulk resistance coefficient decreases and the dielectric loss increases. High humidity also causes chemical alteration of the material through expansion, swelling, and mould growth, thereby accelerating deterioration.

(3) *Deterioration due to cold–hot cycles*: When insulation material is repeatedly heated and cooled, the deterioration will be accelerated by the mechanical strain from expansion and contraction as well as through heat deterioration.

6.5.3 *Temperature rise curve and thermal time constant*

As previously stated, various losses during operation cause temperature rise. The heat caused by losses is emitted through conduction, convection, and radiation from the surface of the motor to its surroundings. Therefore, the temperature rise will continue until the heat generated and that emitted become equal.

Let us now define the following quantities:

Q = total heat generated inside the instrument due to losses,
C = the average capacity of the instrument,
H = heat emission coefficient,
θ = temperature rise,
t = time from start of driving.

If the temperature has risen by $d\theta$ over a short time dt, then the amount of heat generated within the motor during this time period is $Q\,dt$. $Q\,dt$ is equal to the sum of the heat used for the temperature rise $C\,d\theta$ and the amount of heat emitted outside during dt seconds is $H\theta\,dt$. Hence we have the equation

$$Q\,dt = C\,d\theta + H\theta\,dt \qquad (6.55)$$

(Heat generated = Heat stored + Heat emitted).

This is a differential equation with respect to t and can be solved as

$$\theta = A\,e^{-(H/c)t} + \frac{Q}{H}. \qquad (6.56)$$

Since $\theta = 0$ at $t = 0$, the integration constant A is given as

$$A = -\frac{H}{Q}. \qquad (6.57)$$

Therefore, the temperature rise at a given time t is

$$\theta = \frac{Q}{H}(1 - e^{-(H/c)t}). \qquad (6.58)$$

Replacing the final temperature rise Q/H with θ_m, and the thermal constant C/H with T_θ, we obtain

$$\theta = \theta_m(1 - e^{-t/T_\theta}). \qquad (6.59)$$

This is the fundamental equation which determines the temperature rise, and Fig. 6.8 illustrates the curve for this function.

6.5.4   *Note on measurements*

Data required for the calculation of static characteristics are obtained from the no-load and lock-loaded test. Here, a simple measuring method for each piece of data is described.

(1) *Voltage and current*   Figure 6.9 shows two different circuits. Either of these circuits can be used depending on the type of measurement required. Circuit (a) should be used for the no-load or a light-load test

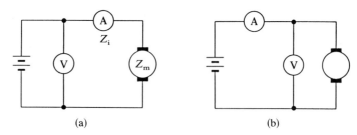

(a)   (b)

Fig. 6.9. Circuits for measuring parameters: (a) no-load or light-load test, (b) lock-loaded test.

(refer to Fig. 6.2), since the impedance Z_m which is seen at the motor terminals far exceeds the impedance of the ammeter Z_i during rotation; the voltage drop across the ammeter is negligible. If circuit (b) is used for the no-load or light-load test, the ammeter indicates the sum of the currents through the voltmeter and the motor, since Z_m and Z_v, the impedance of the voltmeter, are comparable.

Circuit (b) should be used for a lock-loaded test because Z_m (only the armature resistance R_a when speed is zero) is much smaller than Z_v. Here, the current through the voltmeter would be negligible. If circuit (a) is used for a lock-load test, the voltmeter measures the voltage across $Z_i + Z_m$ (the ammeter and motor), producing a large error.

Dealing with the data obtained from the lock-loaded test is explained in Section 6.4.1.

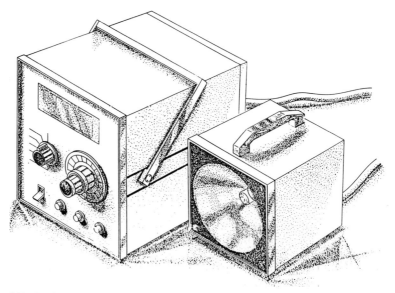

Fig. 6.10. Stroboscope.

(2) *Rotational speed* There are several instruments capable of measuring the rotational speed of a motor. For a small electric motor, a stroboscope (shown in Fig. 6.10) is the most appropriate since the speed can be measured without direct contact with the motor shaft.

(3) *Torque* There are several means of applying load torques in measuring motor torques. The most basic principle is the prony (or cord and pulley) brake. The load torques, which balance with the motor torques, are measured by one of two spring scales as illustrated in Fig. 6.11(a) and (b). An explanation of these follows:

(a) *Two-scale method.* Let the radius of the pulley be R(m), the reading on the left scale W_2(kg), and the reading on the right scale

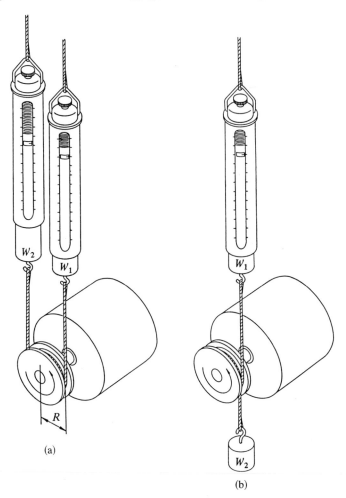

(a)

(b)

Fig. 6.11. Measuring motor torques by the use of spring scales: (a) two-scale method, and (b) one-scale method.

W_1(kg), then the torque T is given as

$$T = 9.8(W_2 - W_1)R \quad \text{(N m)}. \qquad (6.60)$$

(b) *One-scale method.* Let the radius of the pulley be R(m), the reading on the scale W_1(kg), and let the weight be W_2(kg), then the torque T is given as

$$T = 9.8(W_2 - W_1)R \quad \text{(N m)}. \qquad (6.61)$$

This is the same as eqn (6.60).

Reference for Chapter 6

[1] Mas, J. A. (1977). New tool for evaluating PM motors. *Machine Design* **49**, (27), 98–100.

7. Dynamic characteristics of DC motors

In the previous chapter we discussed the static characteristics of conventional and brushless DC motors. This chapter will focus on the theory and measurements of the dynamic characteristics of permanent-magnet DC motors. Dynamic characteristics refer to how a motor responds to operational commands. When considering the dynamic characteristics, it is convenient to represent a motor using an equivalent circuit. The derivation of equivalent circuits of conventional DC motors will first be discussed, and second their transfer functions will be defined and examined from various viewpoints. Third, measurements of parameters governing the dynamic behaviour will be discussed, and finally we will see some of the problems concerned with the matching of a motor and its load.

Some knowledge of Laplace transforms is necessary when considering the transfer functions, therefore a brief discussion of the Laplace transforms will be included.

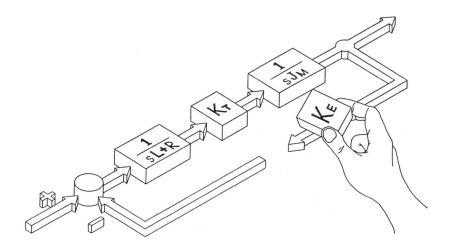

7.1 Equivalent circuits and transfer functions for dynamic characteristics

Let us start with a very basic equivalent circuit. Rotor inertia is the most important factor affecting the dynamic characteristics and it can be expressed as a capacitor in equivalent circuits. The circuit of Fig. 7.1, which is used for explaining the static characteristics, is modified to the equivalent circuit (b) for the analysis of the dynamic behaviour of a DC motor.

The resistance R_D and capacitance C_J in Fig. 7.1(b) have the following meanings:

(1) The kinetic energy $1/2 \cdot J\omega^2$ which is related to the rotor inertia is replaced by the electrostatic energy $1/2 \cdot e^2/C_J$, where C_J is given by

$$C_J = J_M/K^2. \tag{7.1}$$

(2) The frictional heat which is generated when a motor drives a frictional load is replaced in circuit (b) by the Joule loss in resistor R_D, whose value is:

$$R_D = K^2/D \tag{7.2}$$

where J_M is the moment of inertia of the rotor, and K is a motor constant K_T or K_E ($K_T = K_E = K$ because SI units are used here).

In circuit (b) the inductance of the coil is not accounted for.

Mathematical ground of the equivalent circuit When considering the dynamic characteristics, time variants are expressed by the lower case letters listed in Table 7.1.

The torque $m(t)$ produced by the motor is equal to Ki and is used to accelerate the initial load J and to drive the frictional load D.

$$m(t) = Ki = J\frac{d\omega(t)}{dt} + D\omega(t) \tag{7.3}$$

from which we obtain

$$i = \frac{1}{K}\left(J\frac{d\omega(t)}{dt} + D\omega(t)\right). \tag{7.4}$$

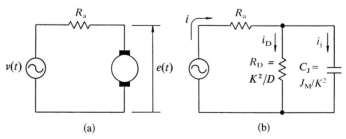

(a)   (b)

Fig. 7.1. Elementary equivalent circuits for analysis of dynamic characteristics; (a) a static equivalent circuit which is a series combination of the armature resistance R_a and back e.m.f. from the armature, and (b) a dynamic equivalent circuit which takes the rotor inertia into consideration.

Table 7.1. List of symbols

	Symbols	Meanings.	SI unit
Variables	$i(t)$, $I(s)$, I	Current	A
	$m(t)$, $T(s)$, T	Torque	N m
	$\omega(t)$, $\Omega(s)$, Ω	Rotational speed	rad s^{-1}
	t	Time	s
	$v(t)$, $V(s)$, V	Terminal voltage	V
	$e(t)$, $E(s)$, E	Back e.m.f. from armature or speed multiplied by K (eqn (1.9))	V
Operator	s	d/dt	

It is assumed that current i splits into i_J and i_D. Then, we obtain

$$i = i_J + i_D = \frac{J}{K}\frac{d\omega(t)}{dt} + \frac{D}{K}\omega(t). \tag{7.5}$$

Since the rotational speed ω and back e.m.f. $e(t)$ are related by

$$e(t) = K\omega(t) \tag{7.6}$$

eqn (7.5) becomes

$$i = i_J + i_D = \frac{J}{K^2}\frac{de(t)}{dt} + \frac{D}{K^2}e(t). \tag{7.7}$$

Therefore,

$$i_J = \frac{J}{K^2}\frac{de(t)}{dt}, \tag{7.8}$$

$$i_D = \frac{D}{K^2}e(t). \tag{7.9}$$

By integrating eqn (7.8) we get

$$e(t) = \frac{K^2}{J}\int i_J\,dt. \tag{7.10}$$

From eqn (7.9) it is seen that the frictional load is represented by a resistance given by eqn (7.2). Also, eqn (7.10) indicates that the inertial load is represented by a capacitance given by eqn (7.1).

The equivalent circuit of Fig. 7.1(b) has the following physical meanings:

(1) The input current multiplied by K is the output torque.

(2) Branch currents i_D and i_J multiplied by K are the torques required to overcome the friction and inertia, respectively.

(3) Capacitor voltage divided by K gives the angular speed.

7.2 Transfer functions

A transfer function is used to evaluate the behaviour of a motor when used in an application where operational conditions change rapidly. This section will first discuss the concept of the transfer function, and then explain in detail the transfer functions of a motor.

7.2.1 *Laplace transforms and transfer functions*

The transfer function is defined as the ratio of the Laplace transform of the output signal to that of the input signal which is applied to a control element. Here, the control element refers to a motor. But many electrical engineers find it convenient to represent an electrical motor by an equivalent circuit. In Fig. 7.2, a motor is compared to an L–R–C circuit. For the motor, the input signal is $v(t)$ and the output is the angular speed $\omega(t)$ and for the circuit, the input signal is $v_i(t)$ and the output is the voltage across the capacitor $v_c(t)$.

We wish to know how the output signal changes with time according to the input signal. However, in many cases, calculations in terms of time t are complicated, and so the functions of t are usually transformed into functions of the complex variable s through a mathematical process called the Laplace transform. On using the example of $v(t)$, the Laplace transform is defined as the integral

$$V(s) = \int_0^\infty v(t)e^{-st}\,dt, \tag{7.11}$$

Also, the Laplace transform $\Omega(s)$ of the angular speed $\omega(t)$ is

$$\Omega(s) = \int_0^\infty \omega(t)e^{-st}\,dt, \tag{7.12}$$

where e is the base of the natural logarithm ($= 2.718$).

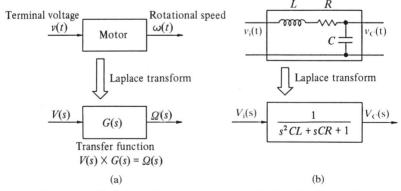

Fig. 7.2. The relationship between input and output signals described in transfer functions: (a) motor, (b) electric circuit.

The following are rules and notes dealing with Laplace transform:

(1) Functions of time t are designated by lower-case letters, and the functions of s by capital letters.

(2) 's' is a complex number which relates to the temporal change of physical quantities.

(3) Because the Laplace transform is an integration of a time-variant function multiplied by the weighting factor e^{-st} over $t = 0$ to $t = \infty$, it includes the characteristics of all temporal changes of the original function.

(4) In most practical cases, it is not necessary to do the calculation of integration, since we can use a table of Laplace transforms.

(5) By convention, all initial values are regarded as being 0 in the Laplace transforms which deal with transfer functions. That is, $v(t)$, $\omega(t)$, $d^n v/dt^n$, and $d^n \omega/dt^n$ are assumed to be 0 at $t = 0$.

(6) For the motor in Fig. 7.2(a), the transfer function $G(s)$ is defined using $V(s)$ and $\Omega(s)$ in eqns (7.11) and (7.12) as

$$G(s) = \frac{\Omega(s)}{V(s)}. \tag{7.13}$$

7.2.2 How to obtain transfer function from circuit diagrams

The voltage equation for the circuit in Fig. 7.2(b) is given as

$$v_i(t) = L\frac{di}{dt} + Ri + \frac{1}{C}\int_0^t i \, dt. \tag{7.14}$$

It is well known that the following equation is obtained after both sides of the above are Laplace transformed:

$$V_i(s) = sLI(s) + RI(s) + \frac{1}{sC} \cdot I(s)$$

$$= \left(sL + R + \frac{1}{sC}\right)I(s) \tag{7.15}$$

where $V_i(s)$ is the Laplace transform of $v_i(t)$ and $I(s)$ is the Laplace transform of the current $i(t)$.

As seen from this example (employing the Laplace transforms of equations, including differentiation and integration) we have the following rules:

$$\text{(differentiation) } d^n/dt^n \text{ can be replaced by } s^n, \tag{7.16}$$

$$\text{(integration) } \int_0^\infty dt \text{ can be replaced by } 1/s. \tag{7.17}$$

From eqn (7.15) the ratio of $V_i(s)$ to $I(s)$ becomes

$$Z(s) = \frac{V_i(s)}{I(s)} = sL + R + \frac{1}{sC} \tag{7.18}$$

Table 7.2. Rules for expressing impedance

	General impedance expressed in terms of Laplace transform	AC impedance $(s = j2\pi f)$ f : frequency	DC impedance $(s = 0)$
Resistor	R	R	R
Coil	sL	$j2\pi fL$	0
Capacitor	$1/sC$	$-j/2\pi fC$	∞

Note. $j = \sqrt{-1}$

which has the dimensions of impedance, and thus is a general impedance expressed in terms of a Laplace transform.

Table 7.2 lists three different concepts of impedances of a resistor, a coil, and a capacitor: (1) general impedance in terms of the Laplace transform, (2) AC impedance, and (3) DC impedance.

To express the relationship between the input voltage v_i and the capacitor v_c by means of the Laplace transform, $V_C(s)$ must first be calculated as follows:

$$V_C(s) = \frac{\text{capacitor's impedance}}{\text{total impedance}} \times \text{input voltage.} \qquad (7.19)$$

That is:

$$V_C(s) = \frac{\dfrac{1}{sC}}{sL + R + \dfrac{1}{sC}} V_i(s). \qquad (7.20)$$

Therefore, the transfer function becomes

$$G(s) = \frac{V_C(s)}{V_i(s)} = \frac{1}{s^2CL + sCR + 1}. \qquad (7.21)$$

As in the above example, if the denominator is second order with respect to s and the numerator is homogeneous, then it is called a second-order transfer function. If the denominator is first order, then it is called a first-order transfer function or the transfer function of the first-order delay.

7.3   Transfer functions of DC motors

Once the equivalent circuit of a motor is known, it is easy to calculate the transfer function of the motor. As seen in the equivalent circuit, Fig. 7.2(b), the rotational speed of the motor is the capacitor voltage divided by the motor constant K. Therefore, the transfer function of the motor $G(s)$ is given by calculating the following formula using the Laplace-transformed impedances:

$$G(s) = \frac{\text{Change in rotational speed}}{\text{Change in terminal voltage}} = \frac{\text{Capacitor voltage}/K}{\text{Terminal voltage}}.$$
(7.22)

Note the words 'change in'. As stated before, when we deal with the transfer functions, we often do not take the initial value of the motor speed into account, i.e. the absolute value of the speed is not considered, but only the changing portions are considered in the transfer function. Next we shall look at some examples of transfer functions.

7.3.1 The transfer function when inductance and friction loads are neglected

When the inductance and friction loads are neglected, the equivalent circuit becomes as simple as Fig. 7.3 and the transfer function is

$$G(s) = \frac{\Omega(s)}{V(s)} = \frac{E(s)/K}{V(s)}$$

$$= \frac{\text{Capacitor's impedance}/K}{\text{Impedance seen from the terminals}}$$

$$= \frac{1/sCK}{R_a + 1/sC} = \frac{1/K}{1 + s(J_M R_a/K^2)}.$$
(7.23)

This is a transfer function of first-order delay.

When the transfer function is of the first-order, and has the form

$$G(s) = \frac{1/K}{1 + s\tau}$$
(7.24)

τ is called the time constant, and this is given by

$$\tau_M = J_M R_a / K^2.$$
(7.25)

Since this is one of the fundamental parameters given in manufacturers' catalogues, let us examine its meaning.

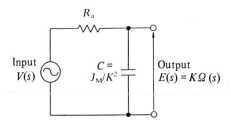

Fig. 7.3. Equivalent circuit neglecting inductance and frictional load.

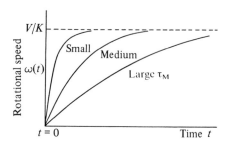

Fig. 7.4. Speed response to a step voltage input applied to a motor having first-order delay transfer function.

If one calculates how the rotational speed $\omega(t)$ changes with time when a voltage V is suddenly applied to a motor whose transfer function is eqn (7.23), it becomes

$$\omega(t) = \frac{V}{K}(1 - e^{-t/\tau_M}). \tag{7.26}$$

When this is plotted (as in Fig. 7.4), it is seen that $\omega(t)$ approaches the final value V/K faster with a smaller τ_M. Thus, the actual response in the rotational speed is delayed as compared with the speed-command voltage. This type of delay is called a first-order delay, and the time required for the speed to become about 63 per cent of the final value is equal to the time constant.

Since it is desirable that time delays be as small as possible in a servosystem, servomotors are required to have a small mechanical time constant. For a motor with a large mechanical time constant, a large gain in the voltage or current amplifier reduces the system time constant and thereby can improve the response. However, such a high gain increases the heat loss generated in the motor and power devices, and can be a cause of unstable operation in the system. When the motor's time constant is small, it is not necessary to increase the gain of the amplifier, which makes the system operation stable and decreases the power loss.

Thus far, we have looked at a terminal voltage that suddenly changes from 0 to V. A similar change is observed when the terminal voltage V_0 of the motor rotating at V_0/K is suddenly changed to $V_0 + \Delta V$.

7.3.2 *The transfer function when the inductance of the armature is not negligible*

When the inductance of the armature is taken into account, the equivalent circuit is as in Fig. 7.5.

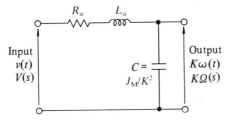

Fig. 7.5. Equivalent circuit considering armature inductance and neglecting frictional load.

In this case, the transfer function is of second order:

$$G(s) = \frac{\Omega(s)}{V(s)} = \frac{\left(\dfrac{K^2}{sJ_M}\right)\left(\dfrac{1}{K}\right)}{R_a + sL_a + K^2/sJ_M}$$

$$= \frac{K}{s^2 L_a J_M + s R_a J_M + K^2}. \tag{7.27}$$

This can be rewritten as

$$G(s) = \frac{\omega_n^2/K}{s^2 + 2\zeta\omega_n s + \omega_n^2} \tag{7.28}$$

where

$$\omega_n \text{ (characteristic angular frequency)} = 1/\sqrt{(\tau_E \tau_M)}, \tag{7.29}$$

$$\zeta \text{ (damping ratio)} = \sqrt{(\tau_M/\tau_E)}, \tag{7.30}$$

$$\tau_E \text{ (electrical time constant)} = L_a/R_a, \tag{7.31}$$

and

$$\tau_M \text{ (mechanical time constant)} = J_M R_a/K^2. \tag{7.32}$$

Figure 7.6 shows how the damping ratio ζ affects the speed-response characteristics for a step voltage applied to the terminals.

7.3.3 Electrical and mechanical time constants

As can be seen from the figure, when the damping ratio ζ is smaller than 1, the response is oscillatory and when the damping ratio $\zeta > 1$, the response is non-oscillatory. Furthermore, when $\zeta \gg 2$, i.e.

$$4\tau_E \ll \tau_M \tag{7.33}$$

the transfer function can be approximated by the following function:

$$G(s) = \frac{1/K}{(s\tau_E + 1)(s\tau_M + 1)} \tag{7.34}$$

and the meanings of electrical and mechanical time constants become clear.

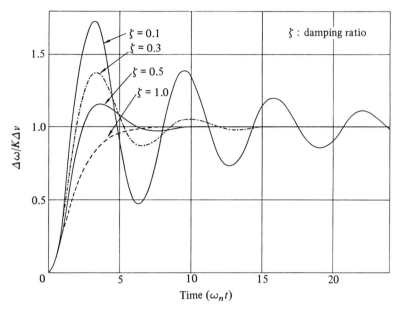

Fig. 7.6. The effect of damping ratios on the angular-speed response for a step voltage applied to the motor terminals.

In many practical cases, since the electrical time constant τ_E is much smaller than the mechanical time constant τ_M, inequality (7.33) applies. Hence the following explanation will be helpful. First the meaning of eqn (7.34) is illustrated in Fig. 7.7. For example, when a step voltage is applied to the motor, a change in the armature current occurs with a delay of the electrical time constant τ_E. Next, in response to this change in the current, the speed increases with a delay of the mechanical time constant τ_M.

If τ_E is extremely small compared with τ_M, it is reasonable to put $\tau_E = 0$.

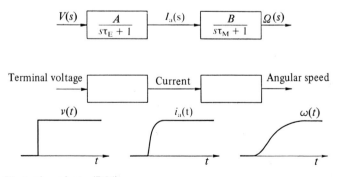

Fig. 7.7. Illustration of eqn (7.34).

The transfer function will be given by eqn (7.24), and the response given by Fig. 7.4.

7.3.4 Electrical and mechanical time constants when a load is applied

Figure 7.8 shows the equivalent circuit when the inertial and frictional loads are added, and the armature inductance is negligible. The transfer function for this case is of first order and given by

$$G(s) = \frac{k}{s\tau + 1} \tag{7.35}$$

where k is the gain constant and τ the time constant which are given as follows:

(1) If the viscous damping coefficient is neglected $(D = 0)$

$$k = 1/K, \tag{7.36}$$

$$\tau = R_a(J_M + J_L)K^2. \tag{7.37}$$

(2) If the viscous damping coefficient is included $(D \neq 0)$

$$k = \frac{K}{K^2 + R_a D}, \tag{7.38}$$

$$\tau = \frac{R_a(J_M + J_L)}{K^2 + R_a D}. \tag{7.39}$$

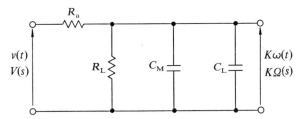

Fig. 7.8. Equivalent circuit considering inertial load, where $C_M = J_M/K^2$, and $C_L = J_L/K^2$.

7.3.5 Transfer functions of higher orders

When an inertial load is coupled to the motor shaft, a mechanical resonance can occur due to the interaction between the shaft/coupling torsion and the load inertia (see Fig. 7.9). The coupling characteristics which produce torsion can be simulated by an appropriate inductance L_s put in the equivalent circuit. Figure 7.10 shows how the rotor and the load are connected with an inductance which represents an elastic coupling, assuming that the load has both inertia and friction.

If the load inertia is J_L, the corresponding capacitance C_L is

$$C_L = J_L/K^2. \tag{7.40}$$

Fig. 7.9. Shaft torsion.

If the viscous damping coefficient of the load is D_L, then the corresponding R_L is

$$R_L = K^2/D_L. \tag{7.41}$$

Let us now consider the magnitude of the inductance which represents the torsion. Suppose torque T is applied to the load with the rotor fixed, and a displacement θ appears as a result. The relation between T and θ may be expressed as

$$T = Q\theta \tag{7.42}$$

where Q is a proportional constant.

Then the equivalent inductance L_s is given as

$$L_s = K^2/Q. \tag{7.43}$$

The proof of eqn (7.43) is as follows.

Proof The electromagnetic energy stored in the inductor and the energy stored in the torsion are considered to be the same:

$$\tfrac{1}{2}L_s i^2 = \tfrac{1}{2}Q\theta^2. \tag{7.44}$$

On the other hand, using eqn (7.42) and the relation $T = K_i$, we can

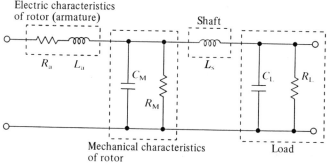

Fig. 7.10. Equivalent circuit representing torsion between load and rotor by inductance L_s.

rewrite the above equation as:

$$\tfrac{1}{2}L_s i^2 = \frac{1}{2}\frac{1}{Q}(Q\theta)^2 = \frac{1}{2}\frac{1}{Q}T^2 = \frac{1}{2}\frac{1}{Q}K^2 i^2. \qquad (7.45)$$

In order for this equation to hold, eqn (7.43) must also hold true.

Transfer function   The transfer function for the equivalent circuit of Fig. 7.10 is written as follows:

$$G(s) = \frac{E(s)/K}{V(s)} = \frac{1}{KF(s)} \cdot \frac{R_L}{1 + sR_L C_L} \qquad (7.46)$$

where $F(s)$ is the impedance of the equivalent circuit as seen from the input side when the output side is open. For reference, $F(s)$ is given as:

$$F(s) = (R_a + sL_a)$$

$$+ \frac{R_M\{sL_s(1 + sR_L C_L) + R_L\}}{(1 + sR_L C_L)(1 + sR_M C_M)sL_s + (R_L + R_M) + s(C_M + C_L)R_M R_L} . \qquad (4.47)$$

Also note that the output terminal voltage $E(s)$ in the equivalent circuit in Fig. 7.10 is the same as the load speed multiplied by K.

7.3.6   *When the influence of brushes is considered*

Thus far we have not considered the voltage drop across the brushes. Figure 7.11 shows an equivalent circuit which considers the brush voltage using two rectifier diodes as shown in Fig. 6.3.

Here, the transfer function can be regarded as being the same as eqn (7.34) or eqn (7.46) while a current is passing through either of the diodes.

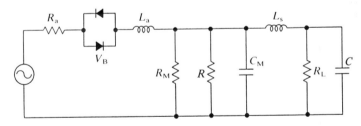

Fig. 7.11. Equivalent circuit considering voltage drop across brushes.

7.3.7   *Transfer function when a driving circuit is included*

For practical servomotor use, we often need to consider a transfer function for the combination of a motor and its drive circuit. We shall now consider two basic examples. Figure 7.12 shows (a) a motor which is driven as the load of an emitter-follower circuit, and (b) a motor which is driven as the load of a constant-current circuit. In either circuit, the input

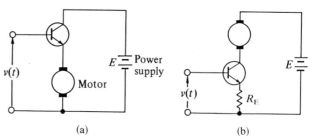

Fig. 7.12. Two basic driving circuits for DC motors: (a) voltage-control type, (b) current-control type.

signal is applied to the base terminal of the transistor. In circuit (a) if the base-emitter forward voltage is neglected, we can use the transfer function which was previously discussed. In circuit (b), however, if the base-emitter voltage is neglected, the emitter current is directly controlled by $v(t)$ as follows:

$$i_E = v/R_E. \tag{7.48}$$

This current is almost the same as the collector current or the motor current.

The transfer function for drive circuit (b) is obtained by the following procedure. First, by substituting $m = K_T i$ into eqn (7.3), we obtain the following equation:

$$K_T i(t) = J\frac{d\omega(t)}{dt} + D\omega(t). \tag{7.49}$$

This is Laplace-transformed into

$$K_T I(s) = (sJ + D)\Omega(s). \tag{7.50}$$

Therefore, the current–speed transfer function is

$$\frac{\Omega(s)}{I(s)} = \frac{K_T}{sJ + D}. \tag{7.51}$$

Finally, by using $V(s) = R_E I(s)$, we obtain the voltage–speed transfer function:

$$\frac{\Omega(s)}{V(s)} = \frac{1}{R_E}\frac{K_T}{sJ + D} \tag{7.52}$$

7.3.8   Transfer function in feedback control system

Figure 7.13 shows two block diagrams of the simplest feedback control systems respectively for voltage control and current control schemes. Here, the speed command is given by V_i and it is compared with $V_0 = \beta\Omega$; the angular speed Ω is regulated so as to minimize the differ-

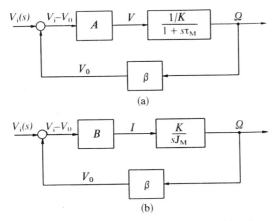

Fig. 7.13. Block diagrams for speed control feedback system: (a) voltage-control type, (b) current-control type.

ence between V_i and V_0. Here, β is a constant of the transducer which converts the rotational speed into voltage information. For simplicity, the motor is simulated by a very simple model in which the armature inductance and the frictional load are neglected. In block diagram (a), 'A' indicates the gain of the amplifier which amplifies the error $(V_1 - V_0)$. Thus the output voltage becomes

$$V = A(V_i - V_0).$$

This is the voltage applied to the motor terminals to drive it. If $V_i > V_0$, the voltage V is large (and positive) and causes the motor to turn faster. When V_0 is slightly lower than V_i, a constant speed specified by V_i will be maintained while the motor receives an appropriate voltage.

In the current control scheme (b), 'B', which has the dimension of conductance, is the gain of the part that amplifies the voltage difference and converts it into the current.

When block diagrams (a) and (b) in Fig. 7.13 are further simplified, both become first-order transfer functions of Fig. 7.14 and their time

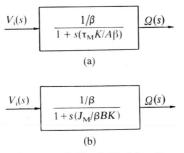

Fig. 7.14. Simplified block diagrams of Fig. 7.13: (a) voltage-control type, (b) current-control type.

constants are

(a) $\tau = (K/A\beta)\tau_M = J_M R_a/\beta AK$ (7.53)

(b) $\tau = J_M/\beta BK$ (7.54)

where $\tau_M = J_M R_a/K^2$ (mechanical time constant of the motor).

The lower these time constants, the quicker the dynamic response and the higher the upper limit frequency at which the speed control is effective. In order to attain low time constants, the mechanical time constant τ_M of the motor itself should be low. One might think that increasing the gains A and B will compensate for a large value of τ_M, but this is not always true. For example, consider the situation where start and stop are repeated. Whenever the command value (input signal) V_i changes, a large value of $V_i - V_0$ appears, and in the current control type, the current I increases considerably so as to accelerate or decelerate the rotor. Because of this high current a lot of power is dissipated into heat inside the armature. A servo-amplifier that allows such a high current must be large in size and so expensive. But if the current is limited, the response will be lessened since there will not be sufficient torque for the acceleration and deceleration. Consequently, it is desirable that the motor has a small mechanical time constant. Necessary conditions for a small mechanical time constant are:

(1) Small moment of inertia J_M.

(2) Large motor constant K.

(3) Low value of R_a.

Proof of the transformation from Fig. 7.13 to 7.14 Here the proof is given for the voltage control system.

$$V_i - V_0 = V_i - \beta\Omega$$ (7.55)

$$V = A(V_i - \beta\Omega)$$ (7.56)

$$\Omega = V \cdot \frac{1/K}{1 + s\tau_M} = \frac{A(V_i - \beta\Omega)/K}{1 + s\tau_M}.$$ (7.57)

From these we obtain

$$\Omega = \frac{A/K}{(1 + A\beta/K + s\tau_M)(K/A\beta)}.$$ (7.58)

We assume here that A is so great that $1 \ll A\beta/K$ is satisfied, and the 1 in the denominator can be neglected. By multiplying both the numerator and denominator by $(K/A\beta)$ we obtain

$$\Omega = \frac{1/\beta}{1 + s(\tau_M K/A\beta)} V.$$ (7.59)

7.4   Gain and phase angle of transfer functions

We often discuss dynamic behaviour of electrical motors using the terms
'gain' and 'phase angle' of a transfer function. Let us study their defini-
tions and meanings for the first- and second-order transfer functions. First
we shall start with the transfer function of eqn (7.34).

$$G(s) = \frac{1/K}{(s\tau_E + 1)(s\tau_M + 1)}. \tag{7.60}$$

7.4.1   DC gain

DC gain means $G(0)$ at $s = 0$. In this example, this gain is $1/K$. Gains are
often expressed in dB. The relation between the absolute value of a gain
and its dB value is:

$$\text{DC gain in dB} = 20 \log (1/K) = -20 \log K. \tag{7.61}$$

Note that the gain is different depending on the system of units used,
i.e. the gain when the SI unit $V \, s \, rad^{-1}$ is used for K, and that when
$V \, k \, r.p.m.^{-1}$ is used are different. The relation between these are given as
follows:

$$\text{Gain in } V \, s \, rad^{-1} = \text{Gain in } V \, k \, r.p.m.^{-1} - 40.4 \, dB$$

$$\text{Gain in } V \, k \, r.p.m.^{-1} = \text{Gain in } V \, s \, rad^{-1} + 40.4 \, dB$$

7.4.2   Frequency transfer functions

When changes in the input signal conform to a sine wave of a frequency f,
we use the transfer function which is derived by replacing the s in eqn
(7.60) with $j2\pi f$.

$$G(j2\pi f) = \frac{1/K}{(1 + j2\pi f\tau_E)(1 + j2\pi f\tau_M)}. \tag{7.62}$$

This is a complex number and its absolute value represents the gain.
Namely,

$$\text{gain } |G(j2\pi f)| = \frac{1/K}{\sqrt{\{1 + (2\pi f\tau_E)^2\}} \cdot \sqrt{\{1 + (2\pi f\tau_M)^2\}}}. \tag{7.63}$$

The phase angle of eqn (7.62) is the delay of the transfer function and
is given by

$$\text{phase angle } \phi = (\tan^{-1} 2\pi f\tau_M) + (\tan^{-1} 2\pi f\tau_E). \tag{7.64}$$

As the frequency of the input signal increases, the gain decreases and
the phase angle increases. This means that the response to the input
signal deteriorates. By putting $\tau_E = 0$ in eqns (7.62) and (7.63), we can
obtain the gain of a motor for the case in which the electrical time
constant is negligible and the motor behaviour is expressed by a first-
order transfer function.

7.5 Measurement of parameters related to dynamic characteristics

Parameters related to dynamic characteristics include the mechanical time constant and the electrical time constant. Here we shall study a practical theory of measuring these parameters.

As seen in Section 7.3.1, when a constant voltage V is suddenly applied to the terminals, the motor speed is governed by eqn (7.26). To obtain the mechanical time constant τ_M based on this theory, one must measure the rotational speed by coupling a tachogenerator to the motor shaft. However, this is not a practical method since a coupling and a tachogenerator must be prepared for each motor.

7.5.1 Theoretical grounds for method of measurement

It is practical to obtain mechanical time constant τ_M from the current behaviour observed immediately after a constant voltage V is suddenly applied to the motor terminals. A theory developed by Page[1] is presented here. Consider the case where an equivalent circuit is given as in Fig. 7.15. This circuit is based on the one in Fig. 7.3, but has been modified so that no-load current can be considered.

The armature current after switch S is closed can be derived from a transient theory; it is

$$i_a = \frac{V}{R_a + R_D} + \frac{VR_D}{R_a(R_a + R_D)} \frac{\tau_M}{(\tau_M - \tau_E)} e^{-t/\tau_M}$$
$$- \frac{V}{R_a}\left(1 + \frac{\tau_E}{\tau_M - \tau_E} \cdot \frac{R_D}{R_a + R_D}\right)e^{-t/\tau_E}. \tag{7.65}$$

There are several approximations which are usable to simplify this equation. If we consider the case in which $\tau_M \gg \tau_E$ is satisfied, eqn (7.65) is approximated by

$$i_a = \frac{V}{R_a + R_D} + \frac{VR_D}{R_a(R_a + R_D)} e^{-t/\tau_M} - \frac{V}{R_a} e^{-t/\tau_E}. \tag{7.66}$$

This equation has the following meanings:
(1) At $t = 0$ the armature current is zero.
(2) After the sudden application of the voltage V, the current increases

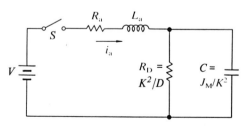

Fig. 7.15. Equivalent circuit suited for measuring dynamic characteristic parameters.

exponentially with the electrical time constant τ_E. For $t < \tau_E$, we have

$$e^{-t/\tau_M} \simeq 1. \tag{7.67}$$

Then eqn (7.66) is approximated by

$$i_a \simeq \frac{V}{R_a} \left(1 - e^{-t/\tau_E} \right). \tag{7.68}$$

When the motor is locked, this equation correctly represents the current rise. Therefore, the electrical time constant can be obtained from this equation.

(3) After time has passed in the range $5\tau_E < t < 5t_M$, we have

$$e^{-t/\tau_E} \simeq 0. \tag{7.69}$$

Therefore, we obtain

$$i_a = \frac{V}{R_a + R_D} + \frac{R_D V}{R_a(R_a + R_D)} e^{-t/\tau_M}. \tag{7.70}$$

Furthermore, if $R_D \gg R_a$, this is further simplified as

$$i_a \simeq \frac{V}{R_D} + \frac{V}{R_a} e^{-t/\tau_M}. \tag{7.71}$$

Since only the mechanical time constant appears in these equations, we can easily measure this value examining this portion of a memory scope trace.

(4) After sufficient time has passed ($t \gg \tau_M$),

$$e^{-t/\tau_E} \simeq e^{-t/\tau_E} \simeq 0. \tag{7.72}$$

Therefore, eqn (7.70) becomes as

$$i_a \simeq \frac{V}{R_a + R_D} \simeq \frac{V}{R_D}. \tag{7.73}$$

This circuit represents the current required to produce the torque required to overcome the friction of the bearings and the brushes.

7.5.2 Method of measurement

The circuit in Fig. 7.16 shows a method of measurement based on the theory presented.[1] The screen on the memory scope displays the voltage and current, and the sweep should be triggered off by the voltage signal. There are two factors to note for this measurement: first, the resistance R_s required to pick up the current should be small compared to the armature resistance R_a of the motor. For example, if R_a is about $2 \sim 3\,\Omega$, then R_s should be about $0.1\,\Omega$. If R_s cannot be made small enough, then compensation is necessary as will be explained later. Secondly, the internal impedance of the power supply is required to be much smaller

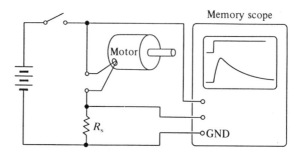

Fig. 7.16. A circuit for measuring dynamic characteristic parameters.

than R_a. This can be verified by the absence of sags in the voltage trace, and if there is a sag, the power supply must be replaced with another.

   Problem 1   When the motor is locked, the current varies with time as shown in Fig. 7.17(a); when the lock is removed, (b) is then obtained. Assuming that $R_a = 1.9\ \Omega$, and $K = 6 \times 10^{-2}\ \mathrm{N\,m\,A^{-1}}$ are already known

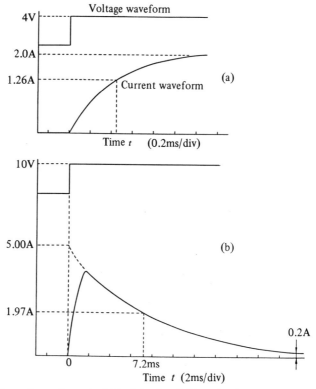

Fig. 7.17. Current waveforms in (a) lock-load and (b) no-load tests.

from static characteristic experiments, calculate τ_E, τ_M and the moment of inertia J_M based on these results. R_s used in the measurement was 0.1 Ω.

Answer and explanations First, from the results of the lock-load test, the stationary current is 2 A. Its 63.2 per cent value is 1.26 A, and the time which is spent before the current reaches this value is 0.46 ms as indicated in Fig. 7.17. Thus, the electrical time constant becomes, with some compensation due to the use of R_s,

$$\tau_E = 0.46 \times (R_a + R_s)/R_a = 0.46 \times 2/1.9 = 0.484 \text{ ms.}$$

Next, because the no-load test is done at a voltage of 10 V, the lock-load current will be

$$10/(1.9 + 0.1) = 5 \text{ A.}$$

On the other hand, since the stationary current is 0.2 A, R_D is

$$R_D = \left(\frac{V}{i_a}\right)_{\text{steady}} - (R_a + R_s) = \frac{10}{0.2} - 2 = 48 \ \Omega.$$

Therefore

$$R_a + R_s + R_D = 50 \ \Omega.$$

The current at $t = \tau_M$ is

$$i_a = V\left\{\frac{1}{R_a + R_D} + \frac{R_D}{R_a(R_a + R_D)}e^{-1}\right\}.$$

Also, the ratio to the lock-load current γ is

$$\gamma = \frac{R_a}{R_a + R_D} + \frac{R_D}{R_a + R_D}e^{-1}$$

$$= \frac{2}{50} + \frac{48}{50} \times \frac{1}{2.718} = 0.3932 = 39.3 \text{ per cent.}$$

The 39.3 per cent of 5 A is 1.97 A, and it takes 7.2 ms for the current to reach this value. τ_M without R_s is compensated for by the following:

$$\tau_M = 7.2 \times \frac{R_a}{R_a + R_s} = 7.2 \times \frac{1.9}{2.0} = 6.84 \text{ ms.}$$

The moment of inertia J_M can be obtained as follows:

$$J_M = \tau_M K^2/R_a$$
$$= 6.84 \times 10^{-3} \times (6 \times 10^{-2})^2/1.9$$
$$= 1.30 \times 10^{-5} \text{ kg m}^2.$$

7.6 Matching motor and load

Of the various types of DC motors, a moving-coil motor having the field magnetic outside the rotor has the smallest mechanical time constant. It is futile to couple a large inertial load directly to this motor, because the advantage of having a small moment of inertia is then totally lost. The most important property of a servomotor is that it is able to accelerate and decelerate quickly. This property should be considered in conjunction with the load to be driven. From this standpoint, we shall now consider some fundamental methods with which to select the motor suitable for a given load.

7.6.1 Power rate

Let us assume that a DC servomotor is accelerating a load through speed-reducing gears as illustrated in Fig. 7.18. The torque needed to drive the load with acceleration $\alpha(=d\omega/dt)$ is at least

$$\text{acceleration torque } T = \left(J_M\xi + \frac{J_L}{\xi}\right)\alpha \qquad (7.74)$$

where ξ is the gear ratio, and α the acceleration of the load or $d\omega_L/dt$.

Since viscous and frictional loads are neglected for simplicity, an equals sign $(=)$ is used in eqn (7.74). The gear efficiency is assumed to be 100 per cent. The moment of inertia of the gear coupled to the rotor is included in the rotor inertia, while that of the gear coupled to the load is included in the load inertia in this theory.

The ξ which minimizes the acceleration of torque T in eqn (7.74) is obtained when the two terms on the right-hand side are equal:

$$\xi = \sqrt{(J_L/J_M)}. \qquad (7.75)$$

By squaring both sides of eqn (7.74), considering the relation of eqn

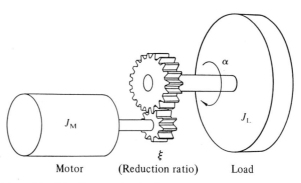

Fig. 7.18. Driving of load by the use of reduction gears.

(7.75), we obtain

$$\frac{T^2}{J_M} = 4\alpha^2 J_L. \tag{7.76}$$

In this equation, the left-hand side is related to the motor and the right-hand side related to the load. When the maximum output torque of the motor T_p is used instead of the torque T on the left-hand side, it is called the power rate:

$$\text{power rate} = T_p^2/J_M. \tag{7.77}$$

Once the moment of inertia of the load J_L and the necessary accelera-tion are determined, the motor can be selected by referring to the power rate in the manufacturers catalogue. An appropriate power rate must be calculated if this data is not found.

Problem 2   According to a catalogue, $J_M = 0.57$ kg cm^2, and the instan-taneous maximum torque $T_p = 5.88$ N m. Calculate the power rate.

Answer and explanations

$$\text{Power rate} = \frac{5.88^2}{5.7 \times 10^{-5}} = 6.07 \times 10^5 \text{ W s}^{-1} = 607 \text{ kW s}^{-1}.$$

However, this value was obtained from the instantaneous maximum torque and it is not desirable to accelerate the motor frequently based on this power rate. When the acceleration is repeated with high frequency, the power rate should be determined using the continuous rated torque. According to the catalogue, the rated torque is 1 N m and this is 1/6 of the instantaneous torque. Therefore, the power rate becomes 1/36 of the previous power rate, and is 18.4 kW s^{-1}.

Problem 3   In the above problem, if the load inertia load is 10^{-4} kg m^2, what is the maximum acceleration?

Answer and explanations   From eqn (7.76), the acceleration α is

$$\alpha = \frac{\sqrt{\text{power rate}}}{2\sqrt{J_L}}.$$

Substituting 607 kW s^{-1} as the power rate into the above equation gives

$$\alpha = \frac{\sqrt{607}}{2 \times \sqrt{10^{-4}}} = \frac{24.6}{2 \times 10^{-2}} = 1.23 \times 10^3 \quad \text{rad s}^{-2}.$$

In other words, this means that the motor can be accelerated from 0 to 1.23×10^3 rad s^{-1} during a second when the gears have a reduction ratio of

$$\xi = \sqrt{J_L/J_M} = \sqrt{10/5.7} = 1.32.$$

This speed in r.p.m. is

$$\frac{1230}{6.28} = 191 \text{ r.p.s.} = 191 \times 60 \text{ r.p.m.} = 11\,460 \text{ r.p.m.}$$

7.6.2   Direct drive

In a magnetic tape system, a capstan is coupled directly to the motor as shown in Fig. 7.19(a), and it is operated in the incremental motion mode. In this type of application, the problem is how to determine the capstan radius R. The following is one way of determining this radius.

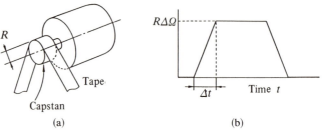

Fig. 7.19. The tape drive: (a) device, (b) speed pattern.

Since this is not a power-rate problem, the problem will be treated differently. When the frictional load is neglected the equation of motion becomes

$$(J_M + J_L) \times \frac{d\omega}{dt} = \text{torque} = KI. \tag{7.78}$$

According to the nature of the application, the tape's acceleration α should be given by

$$\alpha = R\frac{d\omega}{dt} = R\frac{\Delta\Omega}{\Delta t}. \tag{7.79}$$

Substituting this into eqn (7.78), we get

$$(J_M + J_L)\alpha = RKI. \tag{7.80}$$

The moment of inertia is closely related to the radius R, and if the capstan is shaped as in Fig. 7.20(a), the moment of inertia is:

$$J_L = \lambda R^4 \tag{7.81}$$

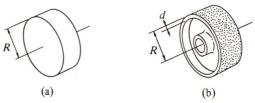

Fig. 7.20. Two basic types of capstan; (a) for a cylindrical capstan the moment of inertia is proportional to R^4, while (b) for a hollow capstan it is proportional to R^3 (where the thickness d does not change while radius R varies).

where λ is a constant dependent on the length and density of the material used. Thus, eqn (7.80) becomes

$$\alpha(J_M + \lambda R^4) = RKI. \qquad (7.82)$$

By dividing both sides by R

$$\alpha\left(\frac{J_M}{R} + \lambda R^3\right) = KI. \qquad (7.83)$$

Assuming that the moment of inertia of the motor (i.e. that of the rotor) J_M is fixed, to obtain the acceleration required by the tape with the least acceleration torque, we must find the R that minimizes the left-hand side of eqn (7.83). Differentiating this as a function of R with respect to R gives:

$$\frac{df(R)}{dR} = \alpha\left(-\frac{J_M}{R^2} + 3\lambda R^2\right). \qquad (7.84)$$

Equating the above to zero gives

$$J_M = 3\lambda R^4 = 3J_L. \qquad (7.85)$$

If a hollow capstan can be used as in Fig. 7.20(b), we have

$$J_L = \lambda' R^3. \qquad (7.86)$$

In this case, the least torque occurs when

$$J_M = 2J_L. \qquad (7.87)$$

As seen from the above, when the load inertia is 1/2–1/3 of the rotor inertia, the radius should be as large as possible.

Reference for Chapter 7

[1] Page, E. B. (1981) An innovative approach to DC motor characterization. *Proc. First motorcon*, 3C2-1–10.

8. Servo-amplifiers and control of DC motors

The movement of a DC motor is controlled by an electronic circuit. The part of such a circuit which is used to drive the motor is called the servo-amplifier. This chapter explains the principles and circuit constructions of various types of servo-amplifiers, and will be followed by a discussion on several methods of speed and position control.

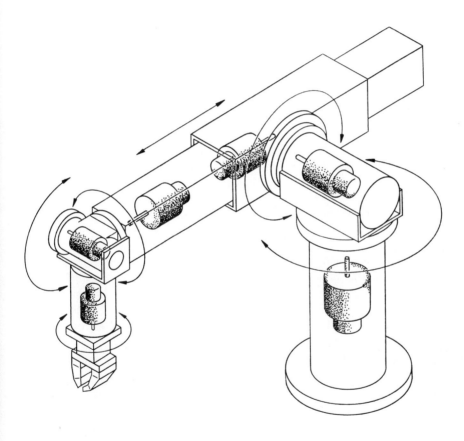

8.1   Basic servo-amplifiers

Servo-amplifiers are classified into two basic categories according to the method they employ to drive the solid-state power devices. One, which is referred to as the linear servo-amplifier, drives bipolar transistors in their linear or active regions; the other is the PWM servo-amplifier which drives bipolar transistors or MOSFETs in the ON–OFF mode, using the Pulse-Width Modulation technique. Both types of servo-amplifier are subsidiarily classified into the voltage-control and current-control schemes.

8.1.1   *Voltage and current controls in linear servo-amplifiers*

With linear servo-amplifiers, we compare the two basic schemes shown in Fig. 8.1 Scheme (a) is a voltage-control type and scheme (b) is a current-control type. Characteristics common to both are:

(1) The input voltage $V_i(t)$, which governs the motor, is applied to the base of the transistor.

(2) The electric power needed to drive the motor is supplied by a power source.

(3) The transistors are driven in the linear or active region.

However, these two schemes are different from each other in terms of what is directly controlled by the transitor. In circuit (a), the motor is driven as the load of an emitter–circuit (a), the motor is driven as the load of an emitter–follower circuit. If the base–emitter voltage, which is about 0.6 V is ignored for simplicity, the input voltage V_i emerges at the motor terminal without a voltage gain. Thus the motor voltage is directly controlled by the input signal. On the other hand the motor current is supplied by the power supply, and its value does not depend on the transistor parameters. The amount of current depends on the applied voltage, speed, and motor parameters. When there is a possibility of transistor damage due to a large current, a current limiter circuit must be added. This is explained later.

On the other hand, in circuit (b), the emitter current i_E is determined

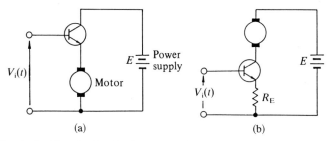

(a)   (b)

Fig. 8.1. Two basic drive circuits of DC motors: (a) voltage-control type, (b) current-control type.

directly by V_i. That is:

$$i_E = \frac{V_i - V_{BE}}{R_E}. \tag{8.1}$$

By neglecting the base–emitter voltage V_{BE}, we can obtain the simple equation:

$$i_E \simeq V_i/R_E. \tag{8.2}$$

This is almost equal to the collector current or the motor current.

In summary, if circuit (b) is used, the motor current is directly controlled by the input voltage V_i. When the transistor is in the linear region, the motor voltage depends on the current, speed, and motor parameters. When the transistor comes into the saturation region, $E - V_i$ is applied to the motor and the current is not given by eqn (8.2).

8.1.2 Voltage-control system

Figure 8.2 shows a voltage-control system having a voltage gain. Three current paths are shown, and the strength of the current is indicated by the thickness of the curves. If the base–emitter forward potential in Tr1 is negligible, the voltage appearing at point Ⓐ, or the voltage across R_A is the same as the input voltage V_i. Note that I_2 is much bigger than I_1 owing to the current amplification in Tr2. Therefore we can ignore the effect of I_1 in the computation of the voltage which appears at point Ⓜ to be applied to the motor. It is V_i multiplied by $(R_A + R_B)/R_A$. When a large current is absorbed by the motor, Tr2 should have a scheme of Darlington connections as shown in Fig. 8.3.

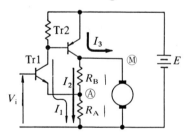

Fig. 8.2. Voltage-controlled servo-amplifier having gain $= (R_A + R_B)/R_A$.

8.1.3 Bipolar drive of voltage controls

The transistor circuits of Fig. 8.1 are for drive in only one direction. To drive a motor in either direction, the circuit in Fig. 8.4 may be used. This circuit uses a PNP transistor and an NPN transitor in a complementary fashion. If we ignore the base–emitter forward potentials in both transistors, the input potential emerges at the emitter of each transistor, and this is applied to the motor. In this circuit the motor current always flows

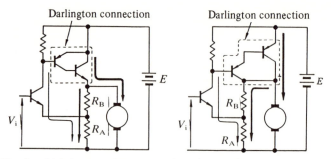

Fig. 8.3. Circuits which increase current capacity using Darlington connections.

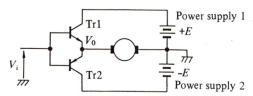

Fig. 8.4. A basic bipolar servo-amplifier of the voltage-control type.

through either Tr1 or Tr2, and the transistors operate in the linear region. For example, when the voltage V_i is positive, Tr1 carries a current which is supplied from the power supply E1. Since the PN junction in this transistor is forward biased, a voltage of approximately 0.6 V appears across the base and emitter. At this time, however, the base–emitter PN junction in Tr2 is reverse-biased and this transistor is in the cutoff region. Conversely, when V_i is negative, Tr2 opens and carries the motor current which is supplied by the power supply E2.

When we consider the base-emitter voltage of each transistor, the relation between V_i and V_0 becomes such as shown in Fig. 8.5(a); a dead zone of about -0.6 to $+0.6$ V appears. Hence when V_i varies as a sine wave, the waveform of V_0 is distorted as shown in Fig. 8.5(b). In Fig. 8.6 this defect is removed by adding two rectifier diodes, and the

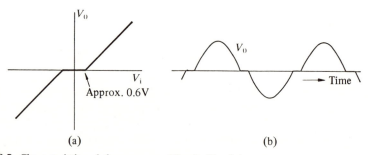

(a)   (b)

Fig. 8.5. Characteristics of the servo-amplifier in Fig. 8.4.

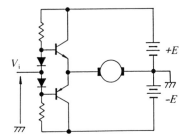

Fig. 8.6. Elimination of the dead zone by using diodes.

influence of the base–emitter junctions in both transistors is compensated for by the PN junctions of the two diodes. However, we cannot expect that the characteristics of these diodes and those of the base–emitter junctions in the transitors are completely identical. Since there is a possibility of current flow in both transistors at the same time, two resistors of low resistance are put between both emitters as shown in Fig. 8.7 in some servo-amplifiers.

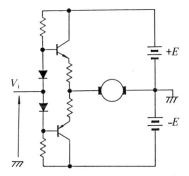

Fig. 8.7. A circuit using low impedance resistors between both emitters.

8.1.4 Bipolar drive having a voltage gain

Figure 8.8 shows a bipolar-drive circuit with a voltage gain of $(R_A + R_B)/R_A$. Tr3 and Tr4 are usually Darlington connected transistors.

8.1.5 Current limiters in the voltage-control schemes

Servo-amplifiers need a current limiting function for the following three reasons:

(1) To avoid demagnetization of the permanent magnets. It is known that demagnetization does not occur in a motor which uses samarium–cobalt magnets. For a motor which uses Alnico or ferrite magnets, there is always a possibility of demagnetization due to a large armature current.

(2) To avoid the destruction of the power transistors.

(3) To avoid burning the armature windings.

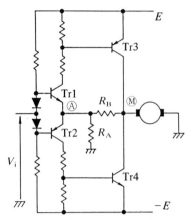

Fig. 8.8. A bipolar driven amplifier having voltage gain $(R_A + R_B)/R_A$.

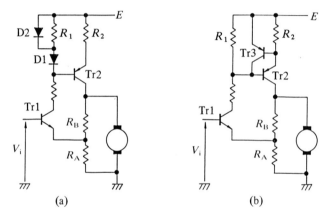

(a)   (b)

Fig. 8.9. Two methods of current limiting in a voltage-control amplifier: (a) using two diodes and a resistor, (b) using a transistor and resistor.

Figure 8.9 shows two examples of current limiters. In circuit (a), a low value resistance R_2 is put between the emitter of Tr2 and the power supply so as to use it as a current sensor. Diode D1 is used here for the compensation of the forward voltage drop between the emitter and base in Tr2. This means that the voltage drop across R_2 is the same as the potential across R_1 or D2. So long as the voltage drop across R_2 is lower than 0.6 V, diode D_2 is never forward biased. When the current through R_2 increases and the voltage across it reaches 0.6 V, diode D2 becomes forward biased and carries a current towards the collector of Tr1. Because of the forward voltage across D2, the voltage drop across R_2 never goes up to 0.6 V. Thus the current will be restricted at $0.6/R_2$ A.

In circuit (b), another transistor Tr3 is used for the current-limiting function. If the voltage drop across R_2 goes over 0.6 V, the emitter–base voltage in Tr3 is forward biased and turns on, and then a current is supplied from the collector of Tr3 to the collector of Tr1. Now Tr3 is on, and this tends to bring Tr2 into the cutoff region, and thus the collector current in Tr2 is limited at $0.6/R_2$ A.

8.1.6 Bipolar amplifier of the current-control type

A basic bipolar linear amplifier of the current-control type is shown in Fig. 8.10. An operational amplifier is used prior to the servo-amplifier for the purpose of controlling the motor current by the input signal V_i. Here, R_s has a low resistance to detect the motor current. The operational amplifier works so as to make the potential at the negative terminal the ground potential. Hence the relation between V_s (the terminal voltage of R_s) and V_i (the input voltage) is

$$V_s = -(R_2/R_1)V_i \qquad (8.3)$$

where $R_s i = V_s$.

Therefore the motor current i is determined by the input voltage V_i:

$$i = -\frac{R_2}{R_1 \cdot R_s} V_i. \qquad (8.4)$$

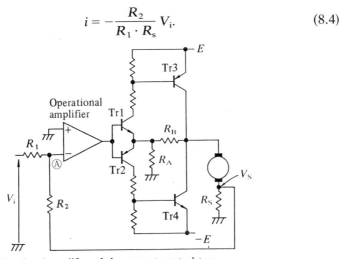

Fig. 8.10. A bi-directional amplifier of the current-control type.

8.2 PWM servo-amplifier

PWM servo-amplifiers are also widely used as drives of small DC motors. In this section we will see the fundamental principles of the PWM drive and its features as compared with linear drives.

8.2.1 Comparison of linear and PWM servo-amplifiers

One feature of linear servo-amplifiers is that their circuits are simple and do not generate harmful electrical noises. However, a lot of power is

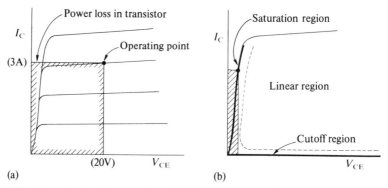

Fig. 8.11. Power loss inside a transistor: (a) linear servo-amplifier dissipates much electrical power into heat, while (b) PWM amplifier produces less heat loss.

dissipated into heat in the final-stage transistors and therefore a large heat sink is needed to remove the heat and protect the transistors from thermal damage. For example, as shown in Fig. 8.11(a), when a voltage of 20 V is applied between the emitter and collector of a power transistor, and it carries a collector current of 3 A, the power loss in it is $3\,A \times 20\,V = 60\,W$.

To reduce the loss in transistors and improve the efficiency of a servo-amplifier, the Pulse-Width Modulation technique should be employed. In a PWM amplifier, a transistor is either in the saturation region or in the cutoff region (see Fig. 8.11(b)). When the transistor is in the saturation region or the fully ON state, the potential between collector and emitter is about 1 V or less, which means the power dissipation is less. When the transitor is in the cutoff region or the fully OFF state, it carries no current, and therefore power loss is absolutely negligible.

8.2.2 Principles of voltage-controlled PWM amplifiers

Figure 8.12 illustrates the basic principle of the PWM amplifier of the voltage-control type. In this circuit, a rectifier diode is connected in parallel with each transistor, and a pulse-width modulator using a comparator IC is put prior to the servo-amplifier. The principle of the modulator is explained in Fig. 8.12(b). The comparator has two input terminals. The voltage control signal is applied to the (+) terminal, and a triangular signal is applied to the (−) terminal. The relationship between the input signal, the output potential on the comparator, and the motor potential are as follows:

(1) When $V_i >$ triangular signal, the output voltage is always equal to $+V_{cc}$, and transistor Tr1 is brought into the fully ON state (saturation region), while Tr2 is in the fully OFF state (cutoff region). Therefore the potential applied to the motor is E.

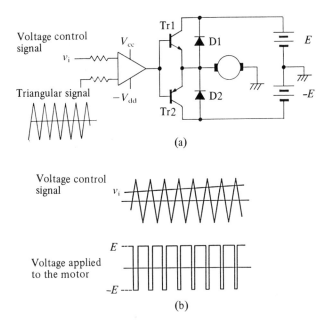

Fig. 8.12. A basic servo-amplifier of the voltage-control type.

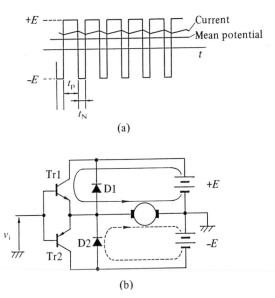

Fig. 8.13. (a) The pulse-width modulated waveform and (b) the mean voltage across the motor driven by circuit.

(2) When $V_i <$ triangular signal, the output voltage is always $-V_{dd}$, and transitor Tr1 is brought into the fully-ON state, while Tr2 in the fully OFF state. The potential applied to the motor is $-E$.

Next, the relationship between the potential waveform and the mean potential across the motor is explained in Fig. 8.13(a). In this figure, the period t_P in which Tr1 is ON is longer than t_N in which Tr2 is ON. Hence the current flows from left to right in the motor as shown by the thin solid curve in Fig. 8.13(b). Since this current is supplied by the $+E$ source, the current polarity is defined as positive. In this period the current increases with time. The current path for the t_N period is shown by the dotted curve in the same figure. The current flows through diode D2 and this is fed back to the power supply $(-E)$. Note that this current decreases with time. When the pulse frequency is high, the next t_P period comes before the current falls to zero. The switching frequency is usually higher than 1 kHz. The higher the frequency, the lower the ripple component in the

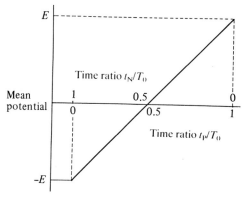

Fig. 8.14. Output voltage of the PWM amplifier shown in the previous figure as a function of time ratio.

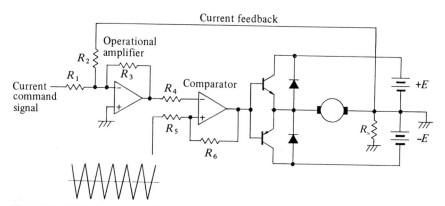

Fig. 8.15. A PWM servo-amplifier of the current-control type.

motor current. Figure 8.14 shows the mean output voltage of the PWM amplifier as a function of the time ratio t_P/T_0.

8.2.3 *Current-controlled PWM amplifiers*

Figure 8.15 is the current-control scheme of the PWM servo-amplifier. Prior to the pulse-width modulator, an operational amplifier was used to control the motor current using a current sensor resistor R_s as explained before in Fig. 8.10.

8.3 Speed control of a DC motor by means of an electronic governor

We shall now look at some motor control systems. Let us start with a speed control known as the electronic governor method.

8.3.1 *Principles of electronic governor controls*

In the steady-state operation of a DC motor, the relation between the terminal voltage V, the armature resistance R_a, the angular speed Ω, and the armature current I_a is given by the following equation.

$$V = R_a I_a + K\Omega \qquad (8.5)$$

where K is the motor constant, which is the same as the torque constant and the back-e.m.f. constant in the SI unit system. Here, we ignore the voltage drop between the brushes and the commutator. In fact, a motor using metal brushes has a negligible voltage drop between brushes and commutator. The voltage drop across graphite brushes will be taken into account later.

From eqn (8.5), we get

$$\Omega = (V - R_a I_a)/K. \qquad (8.6)$$

From this equation we can find a method to control the speed Ω. Since R_a and K are constants, if the motor is set, the speed is changed by the voltage V and the current I_a. Here the voltage V is the quantity which can be easily controlled by external means, and so we can create the method of speed control. At no-load, the current I_a is almost zero, and so the speed is determined solely by the terminal voltage V. When a load is applied to the motor, the current I_a changes corresponding to the load. Hence the speed is not determined solely by the voltage V. For this reason, the current must be detected, and the terminal voltage must be increased so that the speed is kept at the desired value.

8.3.2 *Basic circuit*

As an example, a circuit is shown in Fig. 8.16. The essential points of this circuit are as follows:

(1) The resistances which are used to determine the speed are: R_1, R_2, R_A, R_B, R_s, and the armature resistance R_a. By choosing the proper

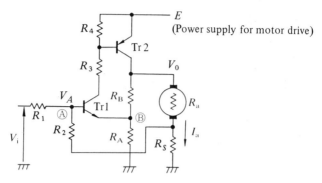

Fig. 8.16. A basic control circuit without using a speed detector.

resistors, we can adjust the speed Ω with the input voltage V_i of the circuit, and make it independent of the current.

(2) When the feedback loop, comprising R_2, is not connected to point Ⓐ, potential V_i appears at point Ⓐ because the current at R_1 is negligible. If the potential difference between the base and the emitter in Tr1 is ignored, the potential at point Ⓑ also equals V_i. In this circuit the current in R_A is a combination of the currents from Tr1 and Tr2. We choose R_3 and R_4 so that the current of Tr1 is below 1/10 of the total current. The voltage V_0, which is applied to the motor, is

$$V_0 \simeq (1 + R_B/R_A) V_i \tag{8.7}$$

where V_0 is the same as V in eqn (8.5).

(3) To adjust the voltage V_0 for compensation of the voltage drop due to the armature current, we need R_s and R_2. R_s is the resistance required to detect the motor current, and R_2 is for feedback of the current information to the input side.

In analysing the circuit with these three points in mind, we can assume that the input impedance of the transitor Tr1 is large enough and that $R_2 \gg R_s$. Hence the electrical potential at point Ⓐ is given by the following equation.

$$V_A = (V_i R_2 + R_s R_1 I_a)/(R_1 + R_2) \tag{8.8}$$

where I_a is the armature current.

If the base–emitter voltage of Tr1 is negligible, this voltage is equal to the voltage at point Ⓑ. Therefore, the terminal voltage V_0 of the motor is this value multiplied by $(R_A + R_B)/R_A$:

$$V_0 = \frac{R_A + R_B}{R_A} \cdot \frac{V_i R_2 + R_s R_1 I_a}{R_1 + R_2}. \tag{8.9}$$

On the other hand, the equation of the motor voltage described in terms of motor parameters is

$$V_0 = (R_a + R_s) I_a + K\Omega. \tag{8.10}$$

By eliminating V_0 from these two equations we obtain

$$K\Omega = \frac{R_2}{R_1+R_2} \cdot \frac{R_A+R_B}{R_A} V_i$$
$$+ \left(\frac{R_1}{R_1+R_2} \cdot \frac{R_A+R_B}{R_A} - \frac{R_a+R_s}{R_s}\right)R_s I_a. \tag{8.11}$$

In order for K to be independent of the current I_a, the following relation is required for the values of the resistances

$$\frac{R_1}{R_1+R_2} \cdot \frac{R_A+R_B}{R_A} = \frac{R_a+R_s}{R_s}. \tag{8.12}$$

Now, the motor speed Ω is proportional to V_i as given by the following equation.

$$\Omega = \frac{R_2}{R_1+R_2} \cdot \frac{R_A+R_B}{R_B}(V_i/K_E). \tag{8.13}$$

If the base–emitter voltage V_{BE}, which is about 0.6 V, and the brush voltage V_B are taken into account, the speed Ω is determined by the following equation

$$K\Omega = \frac{R_2}{R_1+R_2} \cdot \frac{R_A+R_B}{R_A}\left(V_i - \frac{R_1+R_2}{R_2}V_{BE}\right) - V_B. \tag{8.14}$$

(See Section 2.6 about the voltage drop across brushes.)

8.3.3 Example (1)

There are some problems in choosing resistance values in the circuit of Fig. 8.16. When R_1, R_2, R_A, and R_B are determined, the value of R_s should be adjusted so as to satisfy eqn (8.12). For R_s we usually choose a value as low as $1\,\Omega$ or less in order to decrease the power consumption. However, since a variable resistor having such a low resistance is not available, a variable resistor of about $100\,\Omega$ is placed in parallel with this low-valued resistance as in Fig. 8.17, and this is used for the sensitive adjustment of the current. If the ratio of voltage division is ξ, the values of the resistors for this scheme should, instead of eqn (8.12), be determined from

$$\frac{R_1}{R_1+R_2} \cdot \frac{R_A+R_B}{R_A} = \frac{R_a+R_s}{\xi R_s} \tag{8.15}$$

In the circuit of Fig. 8.17, a Zener diode is used to obtain the voltage V_i required for setting the speed. Also the power amplifier uses three transistors in a Darlington connection. The speed control which does not use a speed detector has the advantage of having less parts, but has one drawback; eqn (8.12) can not always be satisfied because R_a and V_{BE} vary with temperature.

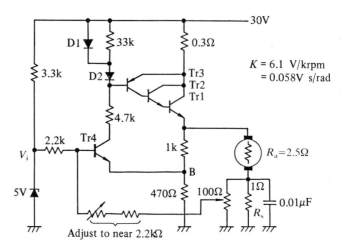

Fig. 8.17. A practical example of a control method without using a speed detector.

8.3.4 *Example* (2)

Another example of the circuit is shown in Fig. 8.18. The control circuit used here is much simpler than the previous one. However, since the main circuit is not capable of starting the motor, a starting circuit is incorporated as an auxiliary circuit. First, let us analyse the operation of the main circuit. Here, the motor uses precious-metal brushes and hence the voltage drop across the brushes is regarded as zero.

The voltage V_A at point Ⓐ, which is the collector terminal of Tr2, is

$$V_A = (R_a + R_s)I + K\Omega. \tag{8.16}$$

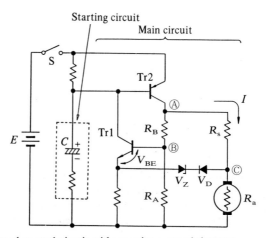

Fig. 8.18. The speed-control circuit without using a speed detector.

Therefore, the voltage V_B at point Ⓑ is

$$V_B = \frac{R_A}{R_A + R_B} \{K\Omega + (R_a + R_s)I\}. \tag{8.17}$$

The emitter voltage of Tr1 is lower than this V_B by V_{BE}, and the voltage V_c at point Ⓒ, which is the terminal voltage of the motor, is higher than this value by the sum of the Zener voltage V_Z, and the diode voltage V_D. Therefore it is given by

$$V_C = \frac{R_A}{R_A + R_B} \{K\Omega + (R_a + R_s)I\} + V_Z + V_D - V_{BE}. \tag{8.18}$$

On the other hand, the motor voltage is given also by the following equation.

$$V_C = K\Omega + R_a I \tag{8.19}$$

assuming that eqns (8.18) and (8.19) are equal, we get

$$\frac{R_A}{R_A + R_B} \{K\Omega + (R_a + R_s)I\} + V_Z + V_D - V_{BE} = K\Omega + R_a I. \tag{8.20}$$

In order that this equation holds independent of current I, the values of R_A, R_B, R_a, and R_s must satisfy the following relation

$$\frac{R_A}{R_B} = \frac{R_a}{R_s}. \tag{8.21}$$

Now the motor speed Ω is determined by

$$\Omega = \frac{R_A + R_B}{KR_B} (V_Z + V_{BE}). \tag{8.22}$$

If the base–emitter voltage V_{BE} of Tr1 and the forward voltage V_D of the diode are equal, the speed Ω becomes

$$\Omega = \frac{R_A + R_B}{KR_B} V_Z. \tag{8.23}$$

It is seen from this equation that the speed is proportional to the Zener voltage V_Z.

When the switch is closed in the circuit shown in Fig. 8.18, a charging current flows through the base of Tr2 to the capacitor C. This current initiates the main circuit. In fact, since the characteristic forward voltage of a diode is independent of current, this is often used instead of an expensive Zener diode as the reference voltage.

8.4 Speed control using a tachogenerator

When the speed is to be regulated across a wide range, a feedback control is employed using a tachogenerator as the speed detector.

8.4.1 *Block diagram and circuit example*

One example of a speed-control circuit using a tachogenerator and its block diagram is shown in Figs. 8.19 and 8.20, respectively. The operation of the circuit is as follows:

(1) The speed-command voltage is given by a potentiometer.

(2) The motor speed is detected by the terminal voltage of the tachogenerator. The relation between the speed Ω and the terminal voltage V_0 is given by

$$V_0 = \beta \Omega \qquad (8.24)$$

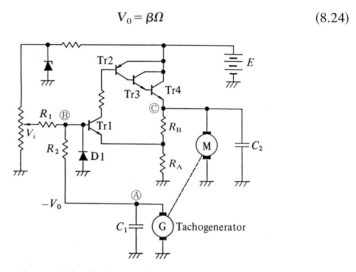

Fig. 8.19. The speed-control circuit using a tachogenerator.

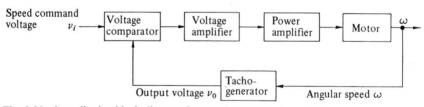

Fig. 8.20. A qualitative block diagram for speed control using a tachogenerator.

where β is the back-e.m.f. constant of the tachogenerator. Here it should be noted that tachogenerator terminals are connected to the circuit so as to generate negative polarity at point Ⓐ with respect to GND. The capacitor connected in parallel with the tachogenerator absorbs the surge voltages which are produced when currents are commutated.

(3) The command voltage V_i and the output voltage V_0 of the tachogenerator are compared through resistors R_1 and R_2—in this case both have the same value R. Hence the voltage at point Ⓑ, the base terminal

of Tr1, is given by

$$V_B = \frac{R_2 V_i - R_1 V_0}{R_1 + R_2} = \tfrac{1}{2}(V_i - V_0). \qquad (8.25)$$

The diode D1 prevents the transistor from being damaged when the voltage at point Ⓑ is negative with respect to the emitter of Tr1.

(4) The voltage given by eqn (8.25) is amplified by the servo-amplifier constructed by Tr1 through Tr4, and emerges at point Ⓒ; the voltage gain being $(R_A + R_B)/R_A$.

(5) The voltage at point Ⓒ is applied to the motor to drive it. The current which is necessary to drive the motor is supplied by the power supply through Tr4.

(6) The capacitor C absorbs the surge voltage generated when the tachogenerator current is commutated.

As mentioned above, when a small voltage error $V_i - V_0$ develops due to a speed decrease less than the required value due to a disturbance, it is amplified to a large voltage and is applied to the motor to increase the speed. This will result in the increse in the output voltage V_0 and reduction of the voltage error $V_i - V_0$, and an appropriate voltage is produced to keep the motor at the regulated speed.

8.4.2 Speed regulation

For the tachogenerator-type control circuit, we need to know how close the actual speed is to the commanded value. To discuss this as a steady-state problem, a quantitative block diagram is presented in Fig. 8.21. Capital letters V_i and V_0, and Ω are used to specify the steady-state voltages and the speed. Now, the load torque T_L is balanced with the output torque T of the motor, and the tachogenerator voltage V_0 is slightly lower than the command voltage V_i.

The voltage error

$$V_i - V_0 = V_i - \beta\Omega \qquad (8.26)$$

is amplified to $A(V_i - \beta\Omega)$ and is applied to the motor. Here, A is the

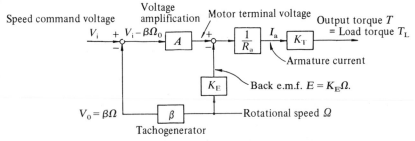

Fig. 8.21. A quantitative block diagram for the steady-state analysis.

gain of the voltage amplification, and its value for the circuit in Fig. 8.19 is

$$A = \frac{1}{2} \cdot \frac{R_A + R_B}{R_A}. \tag{8.27}$$

To determine the motor current, we should subtract the back e.m.f. from the terminal voltage and divide the remainder by the armature resistance R_a. That is

$$I_a = \frac{A(V_i - \beta\Omega) - K_E\Omega}{R_a}. \tag{8.28}$$

Since the torque is I_a multiplied by K_T (the torque constant), and this balances with the load torque, we have

$$A(V_i - \beta\Omega) - K_E\Omega = R_a T_L / K_T. \tag{8.29}$$

This is rewritten as

$$\Omega = \frac{A}{A\beta + K_E} V_i - \frac{R_a T_L}{K_T(A\beta + K_E)} \tag{8.30}$$

If gain A of the amplifier is infinite, the rotational speed is determined by the following equation.

$$\Omega = V_i / \beta. \tag{8.31}$$

This means that the rotational speed Ω is determined by only the command voltage V_i and the constant β of the tachogenerator independent of the external torque T_L.

However, since A is finite in fact, Ω relates to T_L. Therefore if the load torque is increased by ΔT_L with the command voltage V_i kept constant, the speed variation $\Delta\Omega$ is, from eqn (8.30), given by

$$\Delta\Omega = -\frac{R_a}{K_T(A\beta + K_E)} \Delta T_L. \tag{8.32}$$

The ratio of $\Delta\Omega$ to the rated speed Ω_0 is

$$\left|\frac{\Delta\Omega}{\Omega_0}\right| = \frac{R_a \Delta T_L}{K_T(A\beta + K_E)\Omega_0}. \tag{8.33}$$

This is called the speed regulation. Gain G_{L0} of the first-order transfer function, which will be explained later, is

$$G_{L0} = A\beta / K_E. \tag{8.34}$$

Therefore, by using this relation, we can rewrite eqn (8.33) as

$$\left|\frac{\Delta\Omega}{\Omega_0}\right| = \frac{R_a \Delta T_L}{K_T K_E(G_{L0} + 1)\Omega_0}. \tag{8.35}$$

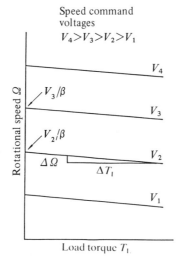

Fig. 8.22. Relations between load torque and rotational speed.

The relation in eqn (8.30) is shown by the graph in Fig. 8.22. Obviously, the lower the driving speed Ω, the larger the speed regulation in eqn (8.33). Therefore, when the speed regulation is specified to be 1 per cent for the rated speed of 2000 r.p.m., the speed regulation for the drive at 500 r.p.m. will be 4 per cent.

In the tachogenerator type speed control system, we can set the control range from 0.1 r.p.m. to some 1000 r.p.m., but we need some means to improve the performance in low-speed ranges.

8.4.3 System time constant and corresponding speed

Figure 8.23 is the quantitative block diagram which is derived from the qualitative block diagram of Fig. 8.20. The value obtained through the multiplication of all the elements shown in the loop in Fig. 8.23 is called the one-loop transfer function, and it is expressed as follows:

$$G_{\mathrm{L}} = \frac{\beta A/K_{\mathrm{E}}}{s\tau_{\mathrm{M}}+1}. \tag{8.36}$$

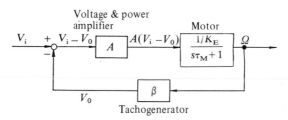

Fig. 8.23. A quantitative block diagram for speed control.

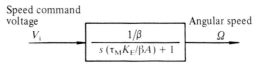

Fig. 8.24. A simplified diagram of Fig. 8.23.

Here, when s is zero, $G_L = G_{L0}$ and this equation becomes the same as eqn (8.34).

Figure 8.24 shows a simple block diagram derived from Fig. 8.23 using an approximation $A/K \gg 1$, which is valid for a large A. Thus, the transfer function of the system is expressed by the first-order-delay form, and its time constant is

$$\tau = \tau_M K_E / A\beta. \tag{8.37}$$

This is not the mechanical time constant which takes account of only the motor, but rather it includes the inertia of the tachogenerator and the load; this is given by the following equation

$$\tau_M = (J_M + J_G + J_L) R_a / K_T K_E \tag{8.38}$$

where J_M is the moment of inertia of the motor, J_G is the moment of inertia of the tachogenerator, and J_L is the moment of inertia of the load.

Equation 8.37 shows that the time constant in the system is inversely proportional to gain A. Therefore, even though the original mechanical time constant τ_M is large, if gain A is also large, the response is high. However, when A is large, even a slight change in speed results in a high voltage which is applied to the motor. This means that we must provide a high-voltage power supply using high-voltage transistors. This makes the cost of the servo-amplifier high. Moreover, since the current flowing in the motor is larger when a high voltage is applied, we need transistors having a large current capacity. Also we must prepare a large heat sink to remove the Joule heat from the servo-amplifier, and some counter-measures should be taken in case of a high temperature rise in the motor.

Hence, when the mechanical time constant of the motor itself is small, and the tachogenerator and the load have a low moment of inertia, a design of a high-response system is possible without making the voltage gain A so large.

8.5 Speed control using a pulse generator

Recently speed-control systems which use a compact pulse generator directly coupled to the rotor have become common. The pulse generator detects the speed by the pulse frequency generated. The block diagram of this system and a circuit are shown in Figs. 8.25 and 8.26, respectively. The pulse generator is the major differentiating factor from the tacho-

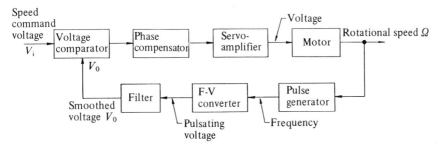

Fig. 8.25. A block diagram for a method using a pulse generator as the speed detector.

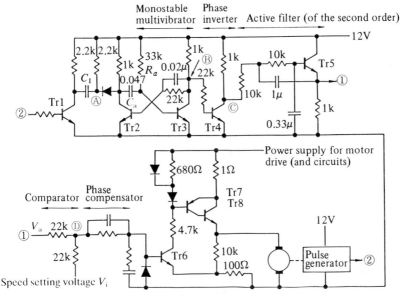

Fig. 8.26. A speed control circuit using a pulse generator as the speed detector.

generator type, so let us start with a more detailed explanation of the pulse generator.

8.5.1 *Pulse generator*

Of all the various types of pulse generators, the magnetic and optical types are the most common. Figure 8.27 illustrates a magnetic type which is known as a kind of AC tachometer. Its rotor is a ferrite magnet having many magnetic poles, and it is usually coupled directly to the motor shaft. An alternating voltage, close to a sine wave, emerges at the stator terminals when the rotor revolves, and it is rectified to a pulse train.

Optical pulse generators, which are commonly known as optical encoders or revolving encoders, are also used. The principles of the low-resolution type and high-resolution type are illustrated in Figs. 8.28(a)

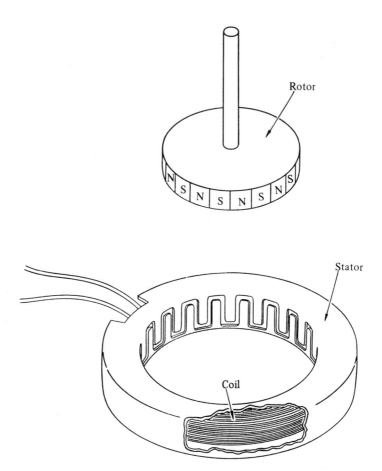

Fig. 8.27. A pulse generator of AC tachogenerator type.

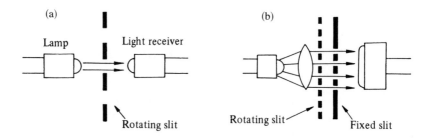

Fig. 8.28. Principles of optical pulse generators: (a) low resolution, (b) high resolution.

Fig. 8.29. Slits used for optical pulse generators: (a) low-resolution type, (b) high-resolution type.

and (b), respectively. Figure 8.29 shows actual examples. Figure 8.29(a) is a low-resolution type made of metal. Because of the small number of slits, however, low-resolution pulse generators can be made by plastic injection moulding. Figure 8.29(b) is a high-resolution type; tiny slits are patterned on a glass or plastic disc. The model shown here has 500 slits.

8.5.2   *Main aspects of circuits*

To understand the outline of the circuit function, we shall look at each wave in Fig. 8.30 and consider its relationship to the block diagram as follows:

(1) The sinusoidal output voltage of the pulse generator is reformed into a square wave using a Schmitt trigger or similar circuit. This is incorporated in the block labelled as 'pulse generator' in Fig. 8.26, and its output terminal is indicated by ②.

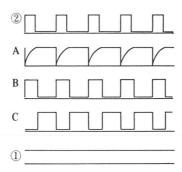

Fig. 8.30. Waveforms in the circuit of Fig. 8.26.

(2) When a high-level pulse is generated, capacitor C_1 is discharged through Tr1 and the voltage at point Ⓐ collapses quickly and becomes the trigger signal to the monostable multivibrator, which is the part made up of Tr2 and Tr3. The voltage at point Ⓑ is close to V_{cc} for a certain period of time, which is approximately $0.69 \times C_a R_a$ s. Therefore, when the motor is revolving at a constant speed and the pulse generator is producing a pulse train at a constant frequency, the waveform at point Ⓑ becomes like Fig. 8.30B. Since Tr4 is used as a phase inverter we have a reversed square wave at point Ⓒ.

(3) By the second-order active filter in the next stage, the alternating component is removed from this signal. The voltage at terminal ① is now a DC voltage which decreases with the pulse frequency. (See Fig. 8.30 ①.)

(4) In the next stage this voltage is compared with the negative speed-command voltage $(-V_i)$ and the difference, which is referred to as voltage error, is detected at point Ⓓ. As explained in Fig. 8.31 the voltage error is $(V_0 - V_i)/2$.

(5) The next stage is the phase compensator circuit to stabilize the system function. In the speed-control system which uses a tachogenerator this is not always needed, but the type we are now examining needs the phase compensation because a filter, which can be a cause of phase delay, is used to eliminate pulsation.

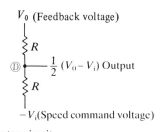

V_0 (Feedback voltage)

R

Ⓓ —— $\frac{1}{2}(V_0 - V_i)$ Output

R

$-V_i$(Speed command voltage)

Fig. 8.31. A voltage comparator circuit.

(6) The final servo-amplifier, which is the part made up of Tr6 through Tr8, amplifies the voltage error by the factor $(R_A + R_B)/R_A$ and sends it to the motor terminal. Here, the current needed by the motor is given by the power supply through Tr7 and Tr8. The details were explained in Section 8.1.

(7) Thus, when the rotational speed decreases and V_0 becomes low, a high voltage is applied to the motor to accelerate it so that V_0 may be close to V_i. Therefore, the rotational speed is automatically regulated at the required speed.

8.6 Position control using a DC motor

The position-control system which uses a DC motor is extensively employed in factory automation equipment and business machines such as printers. We shall see first the basic system which employs an analogue technique. Then a system which uses both analogue and digital electronic circuits will be dealt with.

8.6.1 Basic ideas

Figure 8.32 shows the simplest idea of position control using a DC motor. The position is detected by the output voltage of the potentiometer which is directly connected to the motor or to a geared-down shaft. The position command is given by the voltage. When the output of the potentiometer is lower than the command value, this difference is detected and amplified. The power is then amplified so as to drive the motor. The rotational direction is the direction that brings the tachogenerator's output voltage closer to the position command voltage. The quantitative block diagram of this system is shown in Fig. 8.33. Each part works as follows:

(1) Motor. Because the rotor motion is to be controlled now, we consider the voltage–position transfer function. This is derived by multiplying the voltage–speed transfer function

$$\frac{\theta(s)}{V(s)} = \frac{1/K_E}{s(s\tau_M + 1)(s\tau_E + 1)} \qquad (8.39)$$

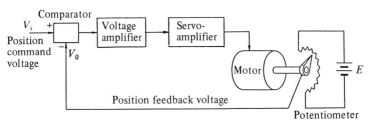

Fig. 8.32. Position control by means of an analogue technique.

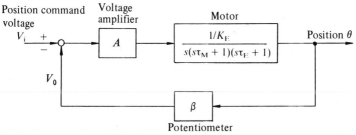

Fig. 8.33. Laplace-transformed block diagram of Fig. 8.32.

by integral operator $1/s$:

$$\frac{\Omega(s)}{V(s)} = \frac{1/K_{\mathrm{E}}}{(s\tau_{\mathrm{M}}+1)(s\tau_{\mathrm{E}}+1)}. \qquad (8.40)$$

This simply means that position is the integral of speed with respect to time.

(2) Potentiometer. In this model the output voltage is zero at the extreme clockwise position, and the voltage increases in proportion to the rotational angle; the constant of proportionality being β.

(3) Voltage and power amplification. The voltage gain in the voltage amplifier is A. Since an emitter follower is used in the servo-amplifier, its potential gain is unity, and so it is not shown in the block diagram in Fig. 8.33.

8.6.2   Stabilization and speed feedback

According to a theory of stability criteria for a feedback-control system, it is well known that this system will be unstable when the voltage gain A is increased. Hence, a phase-compensating element should be incorporated in this circuit to stabilize the system. For this purpose speed feedback is often used. A tachogenerator, which produces a potential proportional to speed, is coupled to the motor and its output voltage is negatively fed back to the input side. A qualitative block diagram for this method is given in Fig. 8.34, and its corresponding quantitative expression in Fig. 8.35. The reason that a negative feedback of speed is effective for stabilization is explained as follows:

First, assume that there is no speed feedback, and the rotor position is approaching the target position. However, position error voltage $V_i - V_0$ has a certain value because the rotor position has not yet reached the target. Since this is amplified and applied to the motor, it is driven towards the target position by a strong torque. Now, because the rotor and load have inertia, the target position is passed. After the target position is exceeded a negative torque is developed in the motor and thus the overshoot is reduced. Next the motor is driven in the opposite direction by this negative torque, and again the target position is ex-

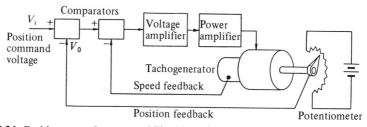

Fig. 8.34. Position-control system stabilized by using a tachogenerator.

Position command voltage

Voltage & power amplifier

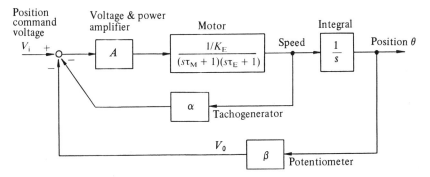

Fig. 8.35. Stabilization effect using speed feedback.

ceeded. Since such overshoots are repeated, an oscillation around the target position occurs. This is known as instability.

Next, we assume that the speed is fed back through the tachogenerator or by some other method. When the rotor position approaches the target position, if the speed is too high even when the rotor comes near the target position, the rotor may overshoot the target position. Therefore, if the speed is detected and is negatively fed back, the motor voltage will decrease. This results in a small overshoot. When the motor returns to the target position, the overshoot is much smaller. Thus the oscillation is damped and motion is stabilized.

8.6.3   An example of a driving circuit

Figure 8.36 shows an example of a circuit. The function of each part is as follows:

(1) Power amplifier. This is a bi-directional servo-amplifier which can

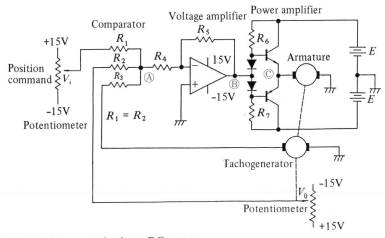

Fig. 8.36. Position control using a DC motor.

drive a motor in either rotational direction. The details of the circuit were explained in Section 8.1.

(2) Comparator. This is composed of only three resistances R_1, R_2, and R_3, where R_1 and R_2 have the same value. It is seen that when the position-command voltage and the output potential from the tacho-generator are equal and opposite, the potential at point Ⓐ is zero, and no potential is applied to the motor.

(3) Voltage amplifier. This is composed of an operational amplifier, and its gain is $-R_5/R_4$.

8.7 Position control using a digital technique

Modern position control of DC motors employs a digital technique to attain an accurate high-resolution positioning. A standard arrangement for a position control servo-mechanism is illustrated in Fig. 8.37. A pulse generator, (PG) which is a device to sense the rotor position and generate output signals, is coupled to the rotor. The pulse generator signals are composed of three channels: A, B, and Zero, as shown in Fig. 8.38. A tachogenerator which is a speed sensor is coupled to the shaft as well.

The error counter is a very important part of this system. This counter is continuously counting the position error, which is defined as the difference between the commanded position and the present position, by means of a digital circuit, and computing the proper value of the speed command as a function of the position error. It is possible to use a microprocessor as the error counter as shown in the photo in Fig. 8.39.

The basic speed–time pattern employed in this system is as shown in Fig. 8.40; it can be divided into three steps: acceleration, slewing, and deceleration.

(1) *Acceleration period.* The motor is accelerated up to the command speed given by the broken line. In this period the velocity servo-loop does

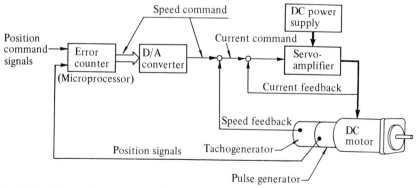

Fig. 8.37. A block diagram of position control using a DC servomotor.

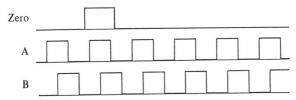

Fig. 8.38. Output signals from the pulse generator; the zero signal is used for the home position, channel A is for counting position and channel B is used for direction discriminator.

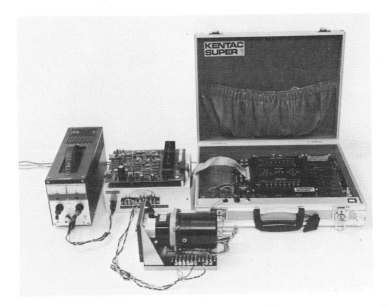

Fig. 8.39. A microcomputer is used as the error counter in a position control system using a DC motor. The microcomputer used here is KENTAC SUPER 1 which was designed for use as a tool for studies of motor controls. (By courtesy of Showa Dengyosha Co., Ltd.)

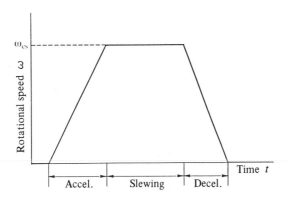

Fig. 8.40. The basic speed–time pattern employed in position control using a DC servomotor.

not function, because the velocity error is so big that the motor current reaches the limit level.

(2) *Slewing.* When the motor speed reaches the command value ω_{cs}, the velocity control comes into effect, and therefore the speed is adjusted at ω_{cs}.

(3) *Deceleration period.* When the position error becomes small, the error counter commands deceleration. The velocity commands are given by

$$\omega_c = \sqrt{(2\delta\varepsilon)} \qquad (8.41)$$

where ω_c is the commanded speed (rad s^{-1})

δ is the decleration rate (rad s^{-2}) and

ε is the position error (rad).

The reason why eqn (8.40) is suitable is as follows, using Fig. 8.41. Now, the motor is decelerated at a rate of δ (rad s^{-2}) at time t. The angular speed ω at this time is given by

$$\omega = (t_0 - t)\delta \qquad (8.42)$$

where t_0 is the time at which the motor will reach the target and stop.

On the other hand, the error E or distance to the target from the present position is the hatched area in Fig. 8.40:

$$\varepsilon = (t_0 - t)\delta(t_0 - t)/2. \qquad (8.43)$$

By eliminating $(t_0 - t)$ from eqns (8.42) and (8.43), we obtain

$$\omega = \sqrt{(2\delta\varepsilon)}. \qquad (8.44)$$

By replacing ω with ω_c, we finally attain eqn (8.41).

The value of B is derived from the equation of motion:

$$\text{Motor torque } T = (J_M + J_L)\delta \qquad (8.45)$$

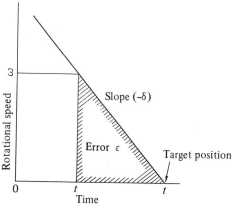

Fig. 8.41. Relation between position error and speed in deceleration.

where J_M is the moment of inertia of the rotor (rotor inertia) and J_L is the moment of inertia of the load.

On the other hand the motor torque T is

$$T = Ki \tag{8.46}$$

where K is the torque constant (motor constant), and i is the motor current, and the maximum value is I.

Therefore the rate of deceleration is given by

$$\delta = TI/(J_M + J_L). \tag{8.47}$$

Thus, δ should be determined according to the load inertia J_L. In a modern servo-amplifier which uses a microprocessor, the load inertia is measured during acceleration, and a suitable rate of deceleration is computed from the acceleration data and other parameters.

(4) *Halt.* When the rotor has reached the target, the speed feedback loop is cut off and the position feedback control is initiated to yield a strong holding torque.

8.8 Incremental-motion control of DC motors

The motional control which repeats start and stop operations is called incremental-motion control. In Fig. 8.42, the sequence of acceleration, slewing, deceleration, and halt are repeated. For slewing periods, the speed is adjusted within a certain constant range as stated in Section 8.7. This type of intermittent drive is widely used in computer terminal units,[1] and one example is the magnetic-tape unit as shown in Fig. 8.43. In the tape, the information is stored in blocks as shown in Fig. 8.44, and is read or written as shown in the same figure. In the slewing period in Fig. 8.44(b) the magnetic head reads or writes the information.

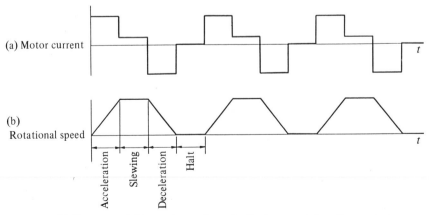

Fig. 8.42. (a) Current and (b) speed waveforms in the incremental motion control.

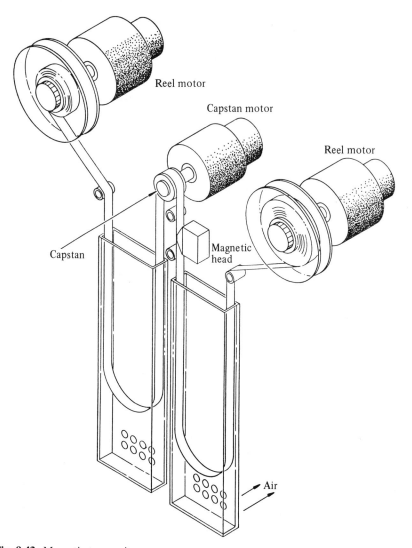

Fig. 8.43. Magnetic tape unit.

In this type of application, as starts and stops need to be done very quickly, a moving-coil motor of the outside field type is used as a capstan motor with forced air-cooling.

During acceleration, the maximum current that the servo-amplifier can supply is fed to the motor. The relation between the angular speed and the current is given by

$$(J_{\mathrm{M}}+J_{\mathrm{L}})\frac{\mathrm{d}\omega}{\mathrm{d}t}=KI, \tag{8.48}$$

where J_{L} is the capstan inertia reflected to the rotor.

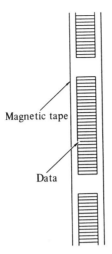

Fig. 8.44. Data stored on magnetic tape.

Reference for Chapter 8

[1] Moritz, F. G. (1977). DC motor stepper systems in computer peripherals. *Proc. Sixth Annual Symposium on Incremental Motion Control Systems and Devices*. Department of Electrical Engineering, University of Illinois, pp. 111–21.

Appendix   Tables of unit conversions

(1) *Weight/Mass*

	Kilogramme (kg)	Pound (lb)	Ounce (oz)
kg	1	2.204 62	35.2739
lb	0.453 592	1	16
oz	0.028 349	0.062 5	1

(2) *Length*

	Metre (m)	Inch (in.)	Foot (ft)
m	1	39.3707	2.380 89
in.	0.025 399	1	0.083 33
ft	0.304 794	12	1

(3) *Torque*

	Newton metre (N m)	Pound inch (lb in.)	Ounce inch (oz in.)
N m	1	8.850 75	141.612
lb in.	0.112 985	1	16
oz in.	0.007 061 55	0.062 5	1

(4) *Rotational speed*

	Radian per second ($rad\,s^{-1}$)	Revolutions per second (r.p.s.)	Revolutions per minute (r.p.m.)
$rad\,s^{-1}$	1	0.159 155	9.549 29
r.p.s.	6.283 19	1	60
r.p.m.	0.104 719	0.016 6667	1

(5) *Moment of inertia*

	$kg\,m^2$	$oz\,in.\,s^2$	$lb\,in.\,s^2$
$kg\,m^2$	1	141.612	8.850 73
$oz\,in.\,s^2$	0.007 061 55	1	0.062 5
$lb\,in.\,s^2$	0.112 985	16	1

Index